Phenolsulfotransferase in
Mental Health
Research

The editorial functions of Dr. Usdin were performed in his private capacity. No official support or endorsement by the National Institute of Mental Health is intended or should be inferred.

Phenolsulfotransferase in Mental Health Research

edited by

MERTON SANDLER

Bernhard Baron Memorial Research Laboratories
and
Institute of Obstetrics and Gynaecology,
Queen Charlotte's Maternity Hospital, London

EARL USDIN

National Institute of Mental Health,
Rockville, Maryland

First published 1981 by
The Scientific and Medical Division
MACMILLAN PUBLISHERS LTD
London and Basingstoke
Companies and representatives throughout the world

ISBN 978-1-349-06120-4   ISBN 978-1-349-06118-1 (eBook)
DOI 10.1007/978-1-349-06118-1

Contents

Preface

Phenolsulfotransferase (PST) (spelled phenolsulphotransferase by our British cousins) is an enzyme whose time is coming rather than one whose time has come. In his introductory chapter, Dr. Sandler gives an overview of the state-of-the-art as well as an historical perspective. But even in the very short period between the time this chapter was written and now, there have been many papers published on PST and several exciting findings have been announced.

A fascinating discussion on sulfation significance was given by Dr. Gerhard Levy at the April 1981 FASEB meeting. The rate of elimination of acetaminophen from plasma is considerably decreased when the administered dose is increased beyond a certain level. The continuing high level of acetaminophen contributes to its toxicity. Dr. Levy hypothesized that the decreased elimination rate resulted from sulfate deficiency; sulfate formation is one of the known metabolic pathways. He administered sulfate at the same time as he administered acetaminophen and, lo and behold, the rate of elimination of even high levels of acetaminophen became the same as that for low levels. Presumably the inorganic sulfate served as a precursor for PAPS which, under PST catalysis, sulfated the acetaminophen. Dr. Levy made the very practical suggestion that addition of sulfate to acetaminophen tablets might lower the number of cases of overdosage toxicity. PST goes marching on!

I shall not insult the reader by summarizing here the contents of the chapters in this volume. He/she can ascertain the general content from the Table of Contents and the details from reading the chapters themselves. Fortunately or unfortunately, the field of PST as applied to mental health research is still sufficiently circumscribed that it is possible to include most of the pertinent material in one volume - as we believe we have done here.

We should like to thank the American College of Neuropsycho-pharmacology (ACNP) for allowing us to have a symposium on PST at the December 1980 meeting of the College; we should like to thank in particular Dr. Oakley Ray and Ms. Terry Bates, without whose help the meeting could not have been held. I need also to give special thanks to my secretary, Ms. Ellen Perella.

And last, but definitely not least, I wish to thank the participants for their participation:

Dr. Ronald T. Borchardt (55-73)
Dept. of Biochemistry, University of Kansas, Lawrence, KS 66044

Dr. Nguyen T. Buu (145-151)
Clinical Research Institute, University of Montreal,
110 Avenue of Pines, West Montreal, Que. H2W 1R7, Canada

Dr. Otto Kuchel (175-185)
Clinical Research Institute, University of Montreal,
110 Avenue of Pines, West Montreal, Que. H2W 1R7, Canada

Dr. Gerard J. Mulder (86-97, 127-144)
Department of Pharmacology, State University of Groningen,
Bloemsingel 1, 9713 BZ Groningen, The Netherlands

Dr. Giovanni B. Picotti (44-54)
Institute of Pharmacology, University of Milan School of
Medicine, Via Manvitelli 32, 20129 Milan, Italy

Dr. Glen Rein (98-126)
Bernhard Baron Mem. Res. Lab., Queen Charlotte's Mat. Hospital
Goldhawk Road, London W6 OXG, U.K.

Dr. Jerome A. Roth (74-85)
Dept. of Pharmacology & Ther., SUNY at Buffalo, Buffalo,
NY 14214

Dr. Merton Sandler (1-7, 186-206)
Bernhard Baron Mem. Res. Lab., Queen Charlotte's Mat. Hospital
Goldhawk Road, London W6 OXG, U.K.

Dr. Nansie S. Sharpless (164-174)
Department of Psychiatry, Albert Einstein College of Medicine
Bronx, NY 10461

Dr. Gertrude M. Tyce (152-163)
Depts. of Physiology & Biochem., Mayo Foundation, Rochester,
MN 55901

Dr. Earl Usdin
   National Inst. of Mental Health, 5600 Fishers Lane, Rm 9C-09
   Rockville, MD 20857

Dr. Godfried M.J. Van Kempen (29-43)
   Biochemical Laboratory, Psychiatric Hospital Endegeest,
   P.O.B. 1250, 2340 BG Oegstgeest, The Netherlands

Dr. Richard M. Weinshilboum (8-28)
   Dept. of Pharmacology & Int. Med., Mayo Foundation/Mayo Clinic
   Rochester, MN 55901

EARL USDIN

Neurosciences Research Branch
National Institute of Mental Health
5600 Fishers Lane, Rockville, MD 20857 USA

Abbreviations

A = adrenaline (cf epinephrine)
ADP = adenosine diphosphate
APS = adenosine 5'-phosphosulfate
ATP = adenosine triphosphate

BSA = bovine serum albumin

CA = catecholamines
COMT = catechol O-methyltransferase
CSF = cerebrospinal fluid

DA = dopamine
$DA-SO_4$ = dopamine sulfate
DBH = DβH = dopamine β-hydroxylase
DBSP = dibromsulphthalein
DCNP = 2,6 dichloro-4-nitrophenol
DHPG = 3,4-dihydroxyphenylglycol
DOMA = 3,4-dihydroxymandelic acid
Dopa = dopa = dihydroxyphenylalanine
DOPAC = dihydroxyphenylacetic acid
DTT = dithiothreitol

E = epinephrine
EDTA = ethylenediamine tetraacetic acid

5-HIAA = 5HIAA = 5-hydroxyindoleacetic acid
HMPG - cf MHPG
HMPE = 4-hydroxy-3-methoxyphenylethanol
5-HT = 5HT = serotonin
HVA = homovanillic acid

LCEC = liquid chrmatogarphy-electrochemical detection

MAO = monoamine oxidase
MHPG = 3-methoxy-4-hydroxyphenylglycol
MN = metanephrine
MOPEG = 3-methoxy-4-hydroxyphenylethyleneglycol
MT = methoxytyramine
MU = 4-methoxyumbelliferone

NMN = normetanephrine

PAEB = procainamide ethobromide
PAP = 3'-phosphoadenosine-5'-phosphate
PAPS = 3'-phosphoadenosine-5'-phosphosulfate
PCA = p-chloroamphetamine
PCP = pentachlorophenol
PP$_i$ = inorganic phosphate
PST = phenolsulfotransferase

RBC = erythrocytes

SAM = SAMe = S adenosylmethionine

VMA = 4-hydroxy-3-methoxymandelic acid

Phenolsulphotransferases: General Introduction

M Sandler

Bernhard Baron Memorial Research Laboratories and Institute of Obstetrics and
Gynaecology, Queen Charlotte's Maternity Hospital, Goldhawk Road,
London W6 0XG, UK

About 100 years ago, Baumann (1876) fed phenol to dogs
and was able to show that some of it was excreted in the urine as
a sulphate conjugate. Since that time, studies of sulphate transfer have pursued a leisurely course and, as far as neuroscience
is concerned, have hardly impinged at all.

The field of sulphatation has a very large and broadly-based
literature (for reviews see, for example, Dodgson and Rose,1970;
Roy,1971; De Meio,1975; Jakoby et al.,1980) which diffuses into
some arcane zoological backwaters (*c.f.* Dodgson,1977). There
seems to be a profusion of sulphotransferases in the animal king-
dom and variations on this enzymatic theme have been identified
in species ranging from primitive single-celled organisms to man.
These enzymes act on a diverse collection of substrates - alkyl
and allyl alcohols (Böstrom and Wengle,1964), phenols (Williams,
1959) and phenolic steroids (Roy,1971), carbohydrates (Dodgson and
Rose,1970) and glycolipids (De Meio,1975). The precise number of
different enzymes, of course, is quite unknown. There are even
different enzymes for individual steroids and separate ones for
heparin and for chondroitin.

These enzymes are found both in the cytoplasm and in the endo-
plasmic reticulum. The rule seems to be that the smaller molecules are degraded by the former group and large-molecule substrates by the latter. They have not been well characterized.
Even the soluble, cytoplasmic enzymes, such as the phenolsulpho-
transferases which form the subject of this present compilation,
are difficult to purify and separate. They tend to aggregate and
change conformation with great facility and, in consequence, present themselves characteristically as multiple peaks on separation
columns. This enzyme or group of enzymes are unstable and difficult to resolve. It is probably wiser, at the present time, to

try to define them in terms of their kinetic characteristics rather than risk the inevitable problems encountered when trying to achieve clear-cut physicochemical separation (De Meio,1975). Even so, more recent studies involving the latter appear to have met with some success (Sekura and Jakoby,1979; Jakoby et al., 1980).

A general scheme of sulphate transfer, starting with inorganic sulphate, is shown below. The progression of events is the same, whatever the chemical nature of the final sulphated product:

$$ATP + SO_4^2 \underset{\longleftarrow}{\overset{ATP\ sulphurylase}{\longrightarrow}} APS + PP_i \qquad (1)$$

$$APS + ATP \underset{\longleftarrow}{\overset{APS\ kinase}{\longrightarrow}} PAPS + ADP \qquad (2)$$

$$PAPS + DA \overset{PST}{\longrightarrow} PAP + DA - SO_4 \quad (3)$$

(1) Inorganic sulphate is activated by ATP to generate adenosine 5'-phosphosulphate (APS). The enzyme carrying out the conversion is ATP-sulphate adenylyl transferase (ATP sulphurylase; EC 2.7.7.5)(Robbins and Lipmann,1958; Robbins, 1958).

(2) APS is further activated by another molecule of ATP to yield 3'-phosphoadenosine 5'-phosphosulphate (PAPS), so called "active sulphate" (Robbins and Lipmann,1957). The enzyme responsible for this step is ATP-adenylyl sulphate 3'-phosphotransferase (APS-kinase; EC 2.7.1.25)(Robbins and Lipmann,1958).

(3) The final step in the series of reactions is transfer of the sulphate group from PAPS to an appropriate acceptor, such as dopamine (DA). In the case of the scheme shown above, the enzyme effecting this conversion is phenolsulphotransferase (PST; EC 2.8.2.1) (Gregory and Lipmann, 1957).

The process of active sulphate formation was studied intensively by Lipmann and his group in the 1950's: this sequence of reactions is likely to have arrived early in the evolutionary scale. It is a relatively primitive mechanism and very expensive in energy.

One of the limiting factors is sulphate availability (Bray et al.,1952) and the presence of large amounts of inorganic or organic sulphate, say in the form of cysteine or cystine, will drive the reaction towards an increased formation of sulphate conjugate (Singer,1975). Conversely, sulphate conjugation may be diminished in the human by the administration of drugs which compete for

available sulphate, the classical example being salicylamide 3
(Levy and Matsuzawa,1967).

Sulphate conjugation of phenols causes a marked change in their
physicochemical and biological properties. Because many of these
conjugates are strongly acidic, it is necessary to store them as
alkaline salts at low temperature in order to minimize autocata-
lytic degradation (Roy and Trudinger,1970).

It was not until 1940 that this enzymatic process began to
appear on the fringes of what is now termed neuroscience when
Richter demonstrated the conjugation of orally-administered adrenaline. PST itself (EC 2.8.2.1) was not identified until the mid-
1950's, by Lipmann's group who at that time called it (incorrect-
ly) "phenolsulphokinase" (Robbins and Lipmann,1957; Gregory and
Lipmann,1957.

PST is widespread in the mammalian body, with the highest specific activity usually being present in the jejunum although this
varies with species. Apart from the presence of this enzyme in the
periphery, it gradually became apparent that some activity must be
present in the brain. Through the 1960's and early 1970's evidence began to accrue that the sulphate conjugates of a variety of
phenolic monoamines and their acidic and alcoholic metabolites
were present in the cerebrospinal fluid, derivatives both of exogenous (Goldstein and Gerber,1963; Schanberg et al.,1968; Taylor
snd Laverty,1969; Goldstein et al.,1970; Sugden and Eccleston,1971)
and then of endogenously (Schanberg et al.,1968a) generated compounds. Their concentrations were studied both with and without the
use of probenecid (Meek and Neff,1973a; Extein et al.,1973; Gordon
et al.,1973). The results were somewhat controversial and there was
some disagreement between the different groups of workers. Finally, in 1972, Hidaka and Austin identified sulphotransferase
activity towards 5-hydroxytryptamine as substrate, whilst the following year Meek and Neff and their group identified PST in discon-
tinuous distribution in rat brain and achieved a limited purification (Meek and Neff,1973b; Foldes and Meek,1973; Meek and Foldes,
1973), and only a matter of months ago did Roth's group further
characterize the enzyme in human brain (Renskers et al.,1980). Its
distribution appears to be somewhat different from that in the rat
(Rein et al. - this volume).

The reader may wonder how one may justify devoting a whole published volume to the study of one rather low profile enzyme or
group of enzymes. The answer to this question arrived suddenly in
1979 when   high PST activity was identified in the human platelet (Hart et al.,1979). If the platelet enzyme proved to be identical to the cerebral version, then we have, as it were, another
window into the brain. We also have a direct approach to possible
changes of the enzyme in human diseases, particularly of the central nervous system. My colleagues and I present data along these

lines within this compilation (Sandler et al. - this volume).

Although there is some degree of overlap, perhaps, in some of the papers presented here, most of the information is quite new. As we shall see, answers are starting to emerge to questions such as whether more than one form of PST exists in man (see Rein et al. - this volume); but there are many questions for which answers are still outstanding. We are not sure whether the platelet enzyme or enzymes are identical with those in the brain. It is not clear whether the platelet enzyme has any local function in the blood stream. Extremely low concentrations only of <u>free</u> dopamine appear to be present in the circulation and this <u>amine</u> seems to be present almost entirely as its sulphate. Might this be a transport form of dopamine and, consequently, is the sulphate always an end-product or may it occasionally be a precursor (Merits,1976; Buu and Kuchel,1979)? Might there be other limiting factors for PST action apart from sulphate availability, e.g. availability of active sulphate or limitations in the sulphate uptake process? Are sulphate conjugates stored as such? What is the nature of endogenous PST inhibitor(s) (Anderson and Weinshilboum,1979)? Because workers in the field have a general impression that the brain enzyme is extremely unstable, how much enzyme is really present there? Apart from the preliminary observations reported within these covers by my colleagues and me (Sandler et al. - this volume),what changes occur in PST activity in human disease? And perhaps the most important question of all; what is the functional role of PST in the human brain? We know that the K_m of PST for an important neurotransmitter like dopamine, for example, is 3 µM (Rein et al.,1981) and that the enzyme is thus avid for its substrate. The K_m of monoamine oxidase for dopamine, however, is somewhat higher, something of the order of 130 µM. However, as the V_{max} for PST is lower in the brain than that of monoamine oxidase by a factor of about 1,000, monoamine oxidase might be presumed to be the more important enzyme for dopamine inactivation (Roth et al. - this volume). We must always remember, however, that local concentrations of activity in particular areas of the brain may be considerably higher and that, in any case, *in vivo* enzymes are not jumbled all together in a homogenate! Membrane barriers and transport systems regulate the access of substrate to particular enzyme active sites. Thus, the raw kinetic data I have quoted are quite insufficient to provide any answer to the question at the present time. From the evidence presented in this volume, the subject seems to have caught fire; it seems likely that many of these questions will soon be answered.

REFERENCES

Anderson,R.J. and Weinshilboum,R.M.(1979). Phenolsulphotransfer-
    ase: enzyme activity and endogenous inhibitors in the human
    erythrocyte. J.Lab.Clin.Med., 94, 158-171.

Baumann,E. (1876). Über gepaarte Schwefelsäuren in Organismus.
    Pflügers Arch.Gen.Physiol.Menschen Thiere 13, 285-308.

Boström,H. and Wengle,B. (1964). Studies on ester sulphates. 19.
    On sulphate conjugation in adult human liver extracts.
    Acta Soc.Med.Upsalien 69, 41-63.

Bray,M.G., Humphris,B.G., Thorpe,W.V., White,K. and Wood,P.B.
    (1952). Kinetic studies of the metabolism of foreign com-
    pounds. 4. The conjugation of phenols with sulphuric acid.
    Biochem.J., 52, 419-423.

Buu,N.T. and Kuchel,O. (1979). The direct conversion of dopamine
    3-0-sulfate to norepinephrine by dopamine-β-hydroxylase.
    Life Sci., 24, 783-790.

De Meio,R.H. (1975). Sulfate activation and transfer. In Metabol-
    ism of Sulfur Compounds, (ed.D.M.Greenberg), Academic Press,
    London, pp.287-358.

Dodgson,K.S. (1977). Conjugation with sulphate. In Drug Metabolism
    from Microbe to Man. (eds. D.V.W.Parke and R.L.Smith), Tay-
    lor & Francis, London, pp.91-104.

Dodgson,K.S. and Rose,F.A. (1970). Sulfoconjugation and sulfohydro-
    lysis. In Metabolic Conjugation and Metabolite Hydrolysis,
    Vol.1 (ed. W.H.Fishman), Academic Press, London, pp.239-325.

Extein,I., Korf,J., Roth,R.H. and Bowers,M.B. (1973). Accumulation
    of 3-methoxy-4-hydroxyphenylglycol sulfate in rabbit cerebro-
    spinal fluid following probenecid. Brain Res., 54, 403-407.

Foldes,A. and Meek,J.L. (1973). Distribution of phenolsulfotrans-
    ferase in rat brain. Fed.Proc., 32, 797.

Goldstein,M.,Anagnoste,B.,Yamamoto,A. and Felch,W.C.Jr. (1970).
    Regional distribution and metabolism of H^3-tyramine in the
    rat brain. J.Pharmacol.Exp.Ther., 171, 196-204.

Goldstein,M. and Gerber,H. (1963). Phenolic alcohols in the brain
    after administration of dopa-C^{14} or dopamine-C^{14}. Life Sci.,
    2, 97-100.

Gordon,E.K., Oliver,J., Goodwin,F.K., Chase,T.N. and Post,R.M.
    (1973). Effect of probenecid on free 3-methoxy-4-hydroxyphen-
    ylethylene glycol (MHPG) and its sulphate in human cerebro-
    spinal fluid. Neuropharmacology 12, 391-396.

6   Gregory,J.D. and Lipmann,F. (1957). The transfer of sulfate among phenolic compounds with 3',5'-diphosphoadenosine as coenzyme. J.Biol.Chem. 229, 1081-1090.

Hart,R.F.,Renskers,K.J.,Nelson,E.B. and Roth,J.A. (1979). Localization and characterization of phenol sulphotransferase in human platelets. Life Sci., 24, 125-130.

Hidaka,H. and Austin,J. (1972). Occurrence and distribution of serotonin-O-sulfotransferase in human brain: a new radio-isotopic assay. Biochim.Biophys.Acta 268, 132-137.

Jakoby,W.B.,Sekura,R.D.,Lyon,E.S.,Marcus,C.J. and Wang,J-L.(1980). Sulfotransferases. In Enzymatic Basis of Detoxication, Vol. 2 (ed.W.B.Jakoby) Academic Press, New York, pp.199-228.

Levy,G. and Matsuzawa,T. (1967). Pharmacokinetics of salicylamide elimination in man. J.Pharmac.Exp.Ther., 156, 285-293.

Meek,J.L. and Foldes,A. (1973). Sulfate conjugates in the brain. In Frontiers in Catecholamine Research. (eds.E.Usdin and S.Snyder) Pergamon Press, New York, pp.167-171.

Meek,J.L. and Neff,N.H. (1972). Acidic and neutral metabolites of norepinephrine: their metabolism and transport from brain. J.Pharmacol.Exp.Ther., 181, 457-462.

Meek,J.L. and Neff,N.H. (1973a). The rate of formation of 3-methoxy-4-hydroxyphenyl ethylene glycol sulfate in brain as an estimate of the rate of formation of norepinephrine. J.Pharmacol.Exp.Ther., 184, 570-575.

Meek,J.L. and Neff,N.H. (1973b). Biogenic amines and their metabolites as substrates for phenol sulphotransferase (EC 2.8. 2.1) of brain and liver. J.Neurochem.,21, 1-10.

Merits,I. (1976). Formation and metabolism of (^{14}C)dopamine 3-O-sulfate in dog, rat and guinea pig. Biochem.Pharmacol., 25, 829-833.

Rein,G.,Glover,V. and Sandler,M. (1981). Sulphate conjugation of biologically active monoamines and their metabolites by human platelet phenolsulphotransferase. Clin.Chim.Acta,in press.

Renskers,K.J.,Feor,K.D. and Roth,J.A. (1980). Sulfation of dopamine and other biogenic amines by human brain phenol sulfotransferase. J.Neurochem., 34, 1362-1368.

Richter,D. (1940). The inactivation of adrenaline in vivo in man. J.Physiol., 98, 361-374.

Robbins,P.W. (1958). Enzymatic synthesis of adenosine-5'-phosphosulfate. J.Biol.Chem.,233, 686-690.

Robbins,P.W. and Lipmann,F. (1957). Isolation and identification of active sulfate. J.Biol.Chem., 229, 837-851.

Robbins,P.W. and Lipmann,F. (1958). Separation of the two enzymatic phases in active sulfate synthesis. J.Biol.Chem., 233, 681-685.

Roy,A.B. (1971). Sulphate conjungation enzymes. In Handbook of Experimental Pharmacology, vol.28 (eds.B.B.Brodie and J.R.Gillette), Springer, Berlin, pp.536-561.

Roy,A.B. and Trudinger,P.A. (1970). The Biochemistry of Inorganic Compounds of Sulphur. Cambridge University Press,Cambridge.

Schanberg,S.M.,Breese,G.R.,Schildkraut,J.J.,Gordon,E.K. and Kopin,I.J. (1968a). 3-Methoxy-4-hydroxyphenylglycol sulfate in brain and cerebrospinal fluid. Biochem.Pharmacol.,17, 2006-2008.

Schanberg,S.M.,Schildkraut,J.J.,Breese,G.R. and Kopin,I.J.(1968b). Metabolism of normetanephrine-H^3 in rat brain - identification of conjugated 3-methoxy-4-hydroxyphenylglycol as the major metabolite. Biochem.Pharmacol.,17, 247-254.

Sekura,R.D. and Jakoby,W.B. (1979). Phenol sulfotransferases. J.Biol.Chem., 254, 5658-5663.

Singer,T.P. (1975). Oxidative metabolism of cysteine and cystine in animal tissues. In Metabolic Pathways: Third Edition,vol.7, (ed.D.M.Greenberg), Academic Press, New York, pp.535-546.

Sugden,R.F. and Eccleston,D.J. (1971). Glycol sulphate ester formation from (^{14}C)noradrenaline in brain and the influence of a COMT inhibitor. J.Neurochem., 18, 2461-2468.

Taylor,K.M. and Laverty,R. (1969). The metabolism of tritiated dopamine in regions of the rat brain in vivo - II. The significance of the neutral metabolites of catecholamines. J.Neurochem.,16, 1367-1376.

Williams,R.T. (1959). Detoxication Mechanisms, 2nd Edition,Chapman & Hall, London.

Phenol Sulphotransferase in Human Platelet and other Tissues: Endogenous Inhibitors, Assay Conditions, Tissue and Substrate Correlations

Richard M Weinshilboum, MD. Robert J Anderson, MD

Clinical Pharmacology Unit, Department of Pharmacology and Internal Medicine,
Mayo Foundation/Mayo Clinic, Rochester, MN 55901 USA

INTRODUCTION

Phenol sulphotransferase (E.C. 2.8.2.1, PST) plays an important role in the sulfate conjugation of catecholamines, catecholamine metabolites, and a variety of drugs such as acetaminophen and alpha-methyldopa (Richter, 1940; Axelrod et al., 1959; Levy et al., 1975; Saavedra et al., 1975; Alam et al., 1977). Individual variation in neurotransmitter function and in response to phenolic and catechol drugs might be due, in part, to individual variations in PST activity. Very little is known about the regulation of PST activity in man. One initial step in the study of individual variation in this important enzyme would be to measure PST activity in an easily obtained peripheral tissue. Since blood is the most easily obtained human tissue, the measurement of PST activity in formed blood elements such as erythrocytes and platelets represents one approach to the study of PST activity in man. However, it cannot be assumed that variations in PST activity in blood elements necessarily reflect significant functional variations in sulfate conjugation. It must first be demonstrated that the biochemical characteristics and the regulation of the enzyme in blood are similar to those in other tissues. Eventually the hypothesis that such variations are of functional significance must also be tested.

A step-wise approach similar to that outlined above has proven useful in the study of other enzymes involved in neuro-transmitter metabolism. For example, human erythrocyte (RBC) catechol-O-methyltransferase activity is under genetic control (Weinshilboum and Raymond, 1977); variation in RBC catechol-O-methyltransferase activity reflects variation of the enzyme activity in the lung, kidney and lymphocyte (Weinshilboum, 1978; Sladek and Weinshilboum, 1980); and variation in the RBC enzyme activity is correlated with functionally significant variation in the methylation of catechol drugs such as L-dopa (Reilly et al., 1980). The series of experiments reviewed below have demonstrated that PST activity is present in the human erythrocyte and platelet. Both human and animal tissues contain potent endogenous PST inhibitors (Anderson and Weinshilboum, 1979), and it is essential that the effects of these inhibitors be negated if the enzyme activity is to be measured accurately. Optimal conditions for the accurate assay of PST activity in human platelet, kidney, and small intestine have been established (Anderson and Weinshilboum, 1980). It has been shown that the human platelet enzyme is capable of catalyzing the sulfate conjugation of a variety of neurotransmitters and drugs and that the relative level of platelet PST is significantly correlated with the enzyme activity in other tissues such as the kidney (Anderson et al., 1980; 1981). These results represent only a first step in the study of PST in man. Additional experiments will be required to determine whether individual variations in PST activity might play an important role in variations in human neurotransmitter function and in drug metabolism.

MATERIALS AND METHODS

Blood Samples

Blood samples were obtained from randomly selected adult white blood donors at the Mayo Clinic Blood Bank. The subjects were unrelated and were not taking any medication. Blood samples were also obtained from 20 adult white patients who underwent clinically indicated nephrectomies for the removal of renal cell carcinomas and from 8 adult white patients who had small bowel biopsy performed. With the exception of the blood used for the purification of PST, samples were collected by venipuncture in 7 ml Vacutainer tubes that contained either powdered sodium EDTA or liquid potassium EDTA. All studies were conducted under the guidelines of the Human Studies Committee of the Mayo Clinic.

Platelet Preparation

Platelets were isolated as described elsewhere (Anderson and Weinshilboum, 1980; Anderson et al., 1981). PST activity was measured in the samples within one hour of the withdrawal of blood.

Kidney and Intestinal Biopsy Sample Preparation

Renal cortical tissue and jejunal mucosal biopsy samples were obtained and prepared as described elsewhere (Anderson and Weinshilboum, 1980; Anderson et al., 1981).

PST Assay

PST activities in human erythrocyte, platelet, kidney,

and gut were measured by the method of Foldes and Meek (1973) as modified by Anderson and Weinshilboum (1979;1980;1981). The assay involved incubation of the enzyme source in the presence of a sulfate acceptor substrate and ^{35}S-3'-phosphoadenosine-5'-phosphosulfate (PAPS), the sulfate donor. In most assays 3-methoxy-4-hydroxyphenylglycol (MHPG) was the sulfate acceptor substrate. This reaction sequence is shown in Figure 1. At the end of the incubation, protein and excess PAPS were precipitated by the addition of barium acetate, barium hydroxide and zinc sulfate. The supernatant remaining after centrifugation was removed, and its radioactivity was determined in a liquid scintillation counter. One unit of enzyme activity represented the formation of 1 nmol of sulfated product per hour of incubation at 37°C. The results were expressed either per mg protein, per 10^8 platelets or per ml of packed erythrocytes.

Protein Determinations

Protein concentrations were measured by the method of Lowry et al. (1951) with bovine serum albumin as a standard.

Purification of Human Erythrocyte PST

Human erythrocyte PST was partially purified from 500 ml of heparinized blood as described in detail elsewhere (Anderson and Weinshilboum, 1979). The purification included ammonium sulfate precipitation of a high-speed supernatant of an erythrocyte lysate followed by gel filtration chromatography performed with Sephadex G-100. The final preparation represented an apparent 415 fold purification from the high-speed supernatant of the initial erythrocyte lysate. The specific activity of the partially purified enzyme was 124 units per mg protein.

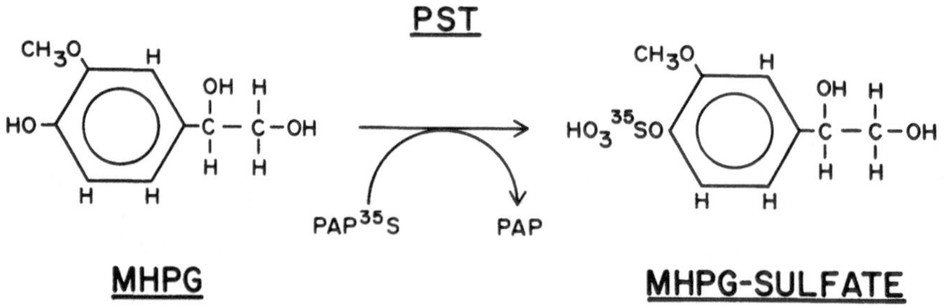

Figure 1. PST reaction sequence. MHPG represents 3-methoxy-4-hydroxyphenylglycol, PAPS represents 3'-phosphoadenosine-5'-phosphosulfate, and PAP represents 3'-phosphoadenosine-5'-phosphate. (Reproduced with permission of Elsevier/North Holland Publishing Company.) (Anderson and Weinshilboum, 1980)

^{35}S-3'-phosphoadenosine-5'-phosphosulfate (0.9-1.5 Ci/mmol) was obtained from New England Nuclear Corporation, Boston, Massachusetts, and was stored at -85°C in aliquots of 100-200 μl. Bis-MHPG piperazine, Tris(hydroxymethyl)aminomethane base, 5-hydroxytryptamine hydrochloride and alpha-methyldopa were purchased from Sigma Chemical Company, St. Louis, Missouri. Tyramine monohydrochloride, 3-hydroxytyramine hydrochloride, and dithiothreitol (Cleland's reagent) were purchased from Calbiochem, San Diego, California. 4'-Hydroxyacetanilide (acetaminophen) was purchased from Eastman Kodak Company, Rochester, New York.

RESULTS

Erythrocyte PST and Endogenous Tissue Inhibitors

RBC lysate PST activity. As a first step in the determination of whether PST activity might be detectable in an easily obtained human cell type, the enzyme activity was assayed in RBC lysates from 178 randomly selected subjects. PST activity was present in every sample. There was a 50 fold intersubject variation in apparent enzyme activity (range from 28-1385 units per ml RBCs with a mean + SEM of 217.7 + 13.1). However, there was not the expected direct relationship between the quantity of RBC lysate and the apparent PST activity (Figure 2). Lack of the expected linear increase in PST activity with increasing quantity of lysate raised the possibility of endogenous enzyme inhibitors. Therefore, RBC PST was partially purified to be used to study possible tissue inhibitors of the enzyme.

Purification and assay of human RBC PST. Human RBC PST activity was partially purified as described under Methods above. This simple purification procedure resulted in an apparent 415 fold purification and an apparent recovery of 193% of the enzyme activity present in a 100,000 x g supernatant. The 193% recovery was further evidence for the existence of endogenous PST inhibitors that were removed during the purification procedure. The approximate molecular weight of human RBC PST estimated from gel filtration chromatography was 40-65,000 daltons. Optimal conditions for the assay of the partially purified enzyme activity were determined, and this enzyme preparation was used to study endogenous tissue inhibitors of PST.

Properties of partially purified human RBC PST. Optimal assay conditions for human RBC PST have been described elsewhere (Anderson and Weinshilboum, 1979). The enzyme had a pH optimum of approximately 7.5 and apparent Km values for MHPG and PAPS of 2.6×10^{-4} M and 4.6×10^{-7} M, respectively. Enzyme

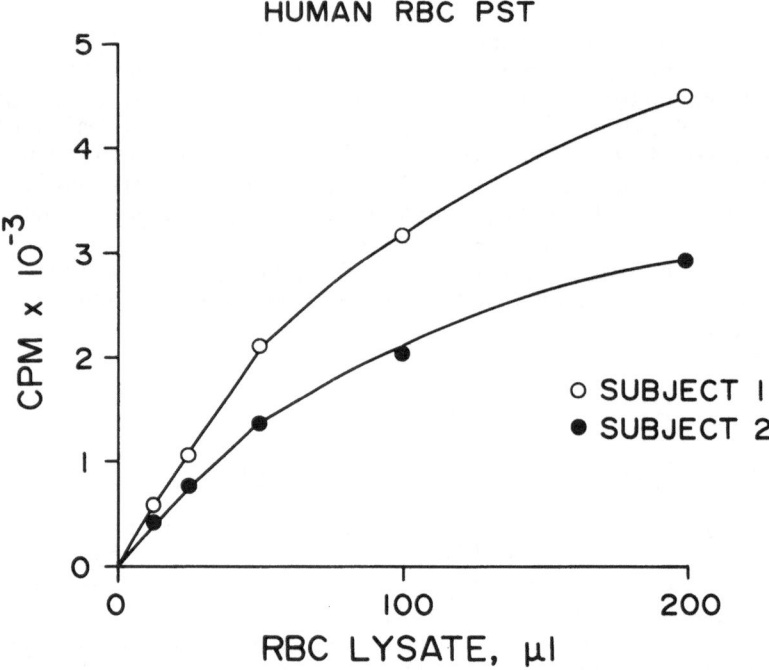

Figure 2. Effect of increasing quantities of RBC lysate on PST activity. Results from two of 178 randomly selected subjects are shown. Each point represents the mean of three determinations. (Reproduced with permission of the C.V. Mosby Publishing Company.) (Anderson and Weinshilboum, 1979)

activity was increased approximately 30% in the presence of 1 mM dithiothreitol, and it was necessary to have 0.25% bovine serum albumin present in the reaction to obtain maximum enzyme activity. The reaction product was shown to be MHPG sulfate by thin layer chromatography. The partially purified RBC enzyme was very sensitive to changes in ionic strength, and enzyme activity decreased in the presence of sodium chloride, potassium chloride, and sodium nitrate at concentrations of greater than 10-20 mM (Anderson and Weinshilboum, 1979).

Endogenous PST inhibitors. To investigate the possibility that erythrocytes contained endogenous inhibitors of PST, partially purified human enzyme was added to each of 20 human RBC lysates (final dilution of RBC contents 1:10, v/v). The average recovery of partially purified enzyme added to lysates was 9.0 + 0.6% (mean + SEM) and varied from 3-15%. A schematic representation of the fashion in which these experiments were performed is shown in Figure 3. These results showed that RBC lysates contained potent PST inhibitors. Variations in apparent PST

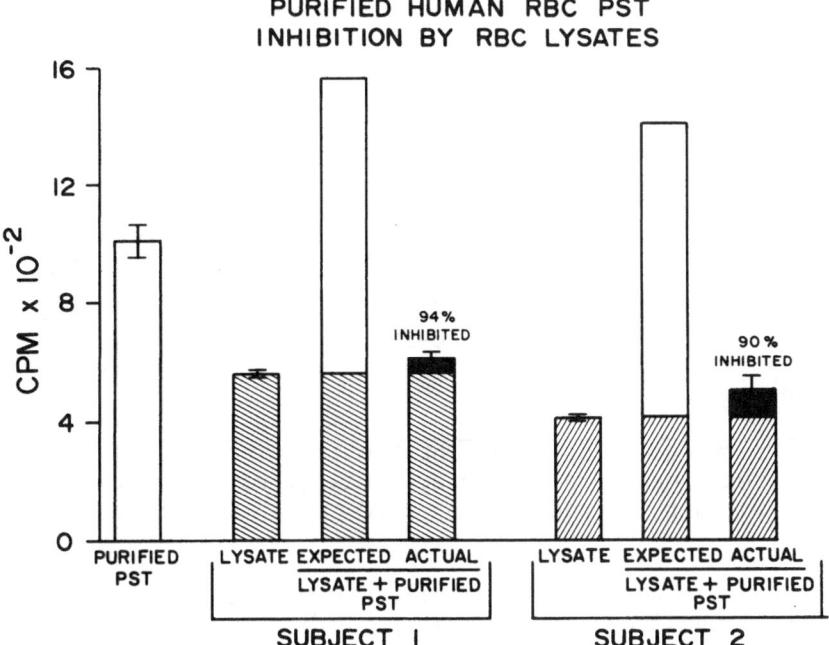

Figure 3. Recovery of purified RBC PST activity added to RBC lysates. Results from 2 of 20 lysates studied are shown. Activities of the purified enzyme, of lysates and of mixtures of the two are shown. Each value represents the mean ± SEM of three determinations.

activities in the 178 RBC lysates described above may have resulted from variations in the content of inhibitors, variations in PST activity, or a combination of the two.

To determine whether human tissues other than erythrocytes contained PST inhibitors, partially purified enzyme was added to each of 12 human renal cortical homogenates (final dilution 1:500), and the enzyme activity was measured. The average recovery of partially purified PST activity from these preparations was only 26 ± 8.1% (mean ± SEM) with a range of 0-83%. Human kidney homogenates also contained potent PST inhibitors.

Finally, homogenates of 6 different rat tissues were prepared and were assayed with and without the addition of partially purified RBC PST. Each of the tissue homogenates inhibited the added enzyme (Table 1). The final dilutions of these tissues were: pituitary 40 fold; brain, kidney, and intestine 500 fold; and liver 5000 fold (Anderson and Weinshilboum, 1979). These experiments demonstrated the presence

Table 1. Rat tissue inhibition of purified human RBC PST

Tissue	Apparent endogenous PST activity (cpm)		Percent inhibition of purified RBC PST	
	Mean ± SEM	Range	Mean	Range
RBC	10+2.4	0-18	94	92-95
Brain	2200+147	1630-2580	30	4-57
Liver	8420+366	7050-9450	82	64-100
Kidney	1870+332	529-2830	58	47-100
Intestine	313+168	0-987	45	22-59
Pituitary (pooled)	987+7	---	68	---

Activities of each tissue (except for the pituitaries) from each of 6 retired breeder female Sprague-Dawley rats were determined with and without the addition of 1660 cpm of RBC PST activity.

of potent inhibitors of PST in dilute rat tissue homogenates. It should be emphasized that in many published studies of PST activity in the rat, tissues that had been diluted only 10 fold were used to measure enzyme activity. PST in those samples was probably highly inhibited.

Characteristics of RBC PST inhibitors. A series of experiments was performed to determine some of the biochemical and physical characteristics of the PST inhibitor(s) in RBC lysates. The results of these experiments have been described in detail elsewhere (Anderson and Weinshilboum, 1979). Thermal stability studies demonstrated that only a portion of the inhibitory activity could be inactivated by heating at 95°C for 10 minutes. The thermostable portion of inhibitory activity was insensitive to treatment with 1 N HCl or 1 N NaOH at 37°C for 30 minutes. In addition, the thermostable inhibitor was not inactivated by exposure to bovine pancreatic alpha-chymotrypsin at 37°C for up to 3 hours. Most of the thermostable inhibitory activity was dialyzable.

Samples of heated and unheated RBC lysate were applied to a Sephadex G-50 gel filtration chromatography column. Both endogenous PST activity and PST inhibitor were measured in each fraction collected. The inhibitor was detected by the addition of partially purified enzyme to an aliquot of each fraction (Figure 4A). Inhibitory activity in the unheated lysate eluted in two major peaks (Figure 4B). The initial inhibitory peak eluted slightly before endogenous PST activity and had an apparent molecular weight of greater than 65,000. A second peak

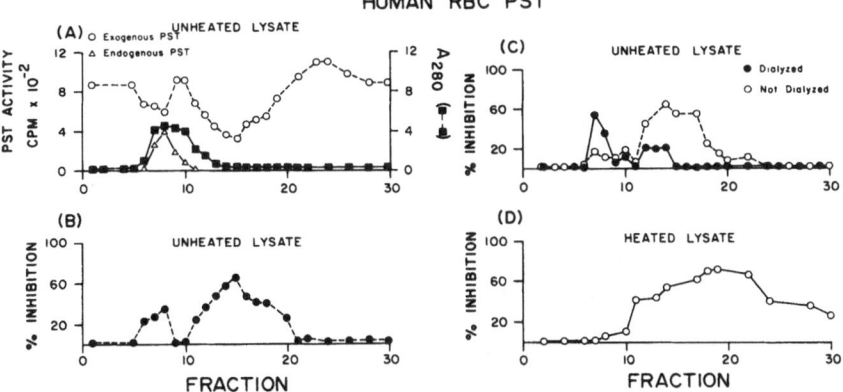

Figure 4. Effect of Sephadex G-50 gel filtration chromatography
on inhibition of PST by unheated and heated RBC lysates. (A)
Activity of endogenous PST and recovery of added exogenous
partially purified RBC PST in each 1.0 ml fraction of column
eluate of unheated RBC lysate. (B) Results from panel A expressed
as percent inhibition. (C) Percent inhibition of PST by the same
unheated lysate after dialysis for 48 hours (●——●) or storage
at 4°C (0----0) prior to gel filtration. (D) Percent inhibition
of PST by the same lysate heated prior to gel filtration.
(Reproduced with the permission of the C.V. Mosby Publishing
Company.) (Anderson and Weinshilboum, 1979)

of inhibitory activity eluted after endogenous PST and had an
estimated molecular weight of less than 2,000. An aliquot of the
same lysate was dialyzed for 48 hours and was applied to the
Sephadex G-50 column. The initial peak of inhibitory activity
remained, but the second peak decreased in size dramatically
(Figure 4C). Conversely, when a heated sample was passed through
the column, the initial high molecular weight inhibitory peak was
absent while the late, low molecular peak remained (Figure 4D).
     These data suggested the presence in human RBC lysates of at
least two endogenous PST inhibitors: a low molecular weight,
thermostable inhibitor and a high molecular weight, thermolabile
inhibitor. The high molecular weight species might represent an
enzyme, the activity of which could result in the destruction of
one of the substrates for the PST reaction. Although such an
enzyme or enzymes would not be a true "inhibitor" of PST, its
presence would result in apparent inhibition.

     Conclusion. These experiments demonstrated that PST
activity was present in the erythrocyte, an easily obtained human
tissue. They also demonstrated that potent endogenous PST
inhibitors were present in the RBC, in the human kidney and in a
variety of rat tissues. There were at least two major classes of
inhibitory species in RBC lysates. One was of relatively low

molecular weight and thermostable and another was of relatively high molecular weight and thermolabile. One important consequence of these experiments was to emphasize the fact that, if variance in PST activity were to be measured accurately, assay procedures that negated the effects of endogenous inhibitors would have to be developed. PST activity was known to be present in the human blood platelet in addition to the activity present in the human erythrocyte (Hart et al., 1979). Therefore, these experiments were expanded to include the determination of optimal conditions for the assay of PST activity in a variety of human tissues including the platelet. In all assays care was taken to negate the effects of endogenous tissue enzyme inhibitors.

Assay of Human Platelet, Kidney and Gut PST Activity

Introduction. Assay procedures which included steps to overcome the effects of endogenous enzyme inhibitors were developed for the measurement of PST activity in the human platelet, kidney, and jejunum (Anderson and Weinshilboum, 1980). After a variety of approaches were tested, extreme dilution was used to negate inhibition in human tissue homogenates. Final assay conditions involved a 10,000 fold dilution of human kidney homogenate, a 400,000 fold dilution of human intestinal mucosa homogenate, and a 160 fold dilution of platelet preparations that were already highly "dilute" since a very small pellet had been homogenized in a 2 ml volume. This high degree of dilution was possible in part because of the availability of very high specific activity ^{35}S-PAPS.

Linearity of enzyme reaction with time and tissue concentration. There was a linear increase in enzyme activity up to 60 minutes of incubation with all three tissues, human platelet, human renal cortex, and human intestinal mucosa. A 30 minute incubation time was used in all assays. PST activity also increased in a linear fashion with increasing quantities of tissue (Figure 5). The amounts of homogenate protein used routinely in the assay procedures were 0.5 to 3.0 µg for platelet, 1.27 µg for kidney and 0.03 µg for gut. The linear relationship between PST activity and tissue concentration was an important point. It was the lack of just such a linear relationship that originally made it clear that the results of assay procedures performed with relatively concentrated tissue homogenates were inaccurate because of the presence of potent tissue inhibitors.

Effect of pH. The effect of pH on PST activity in homogenates of all three tissues was measured at 20°C in the presence of the entire reaction mixture. A pH optimum of approximately 6.2 to 6.8 was found with three different buffer systems, potassium phosphate, Tris HCl, and sodium acetate. The final pH

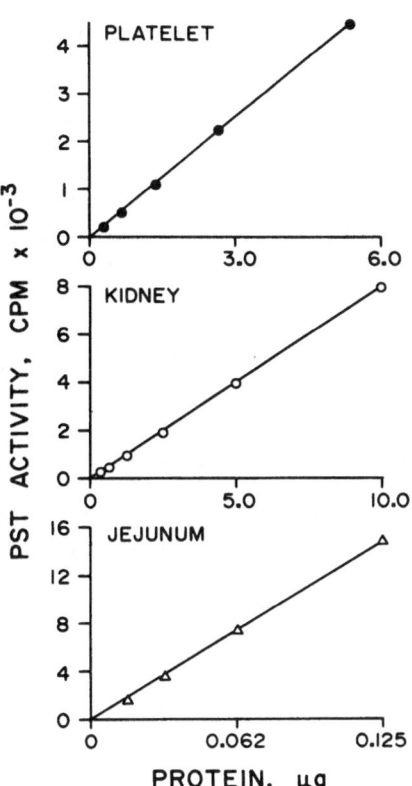

Figure 5. Effect of increasing quantities of tissue on PST activity. Each point represents the mean of three determinations. (Reproduced with the permission of Elsevier/ North Holland Publishing Company.) (Anderson and Weinshilboum, 1980)

of the reaction mixtures under the assay conditions used was 6.5 in all cases. The relatively low ionic strength of the buffer was required because PST is inhibited in the presence of higher ionic strengths (Anderson and Weinshilboum, 1979).

Relationship of enzyme activity to substrate concentration. PST activity in homogenates of all three tissues was measured in the presence of varying concentrations of MHPG and PAPS, the two co-substrates for the reaction. The apparent Michaelis-Menten (Km) constants for MHPG with platelet, kidney, and gut homogenates were 1.09, 0.46, and 1.16 mM, respectively, when determined with the method of Wilkinson (1961) and 1.15, 0.62, and 1.10 mM when estimated by the method of Eisenthal and

Cornish-Bowden (1974). Apparent Km values for PAPS with platelet, kidney, and gut homogenates were 0.14, 0.13, and 0.21 µM, respectively, when determined with the method of Wilkinson (1961) and 0.15, 0.12, and 0.20 µM with the method of Eisenthal and Cornish-Bowden (1974). There were two reasons why it was necessary to use nonsaturating concentrations of PAPS (0.4 µM) in the assay. First, higher concentrations produced striking increases in the values of blank samples that resulted in a decrease in the sensitivity of the assay. Second, it was not economically possible to use higher concentrations of ^{35}S-PAPS. Therefore, the values of PST activity reported here can only be compared with results obtained with similar concentrations of PAPS.

Effects of dithiothreitol and bovine serum albumin. Because the activity of partially purified human erythrocyte PST was increased by the addition of dithiothreitol and bovine serum albumin (BSA), human platelet, kidney, and gut enzyme activities were measured in the presence of various concentrations of these two substances. A final dithiothreitol concentration of 8 mM resulted in a 20-100% increase in PST activity in all three tissues. This concentration of dithiothreitol was used in the reaction mixture. The PST activity of the three tissues increased 80-150% in the presence

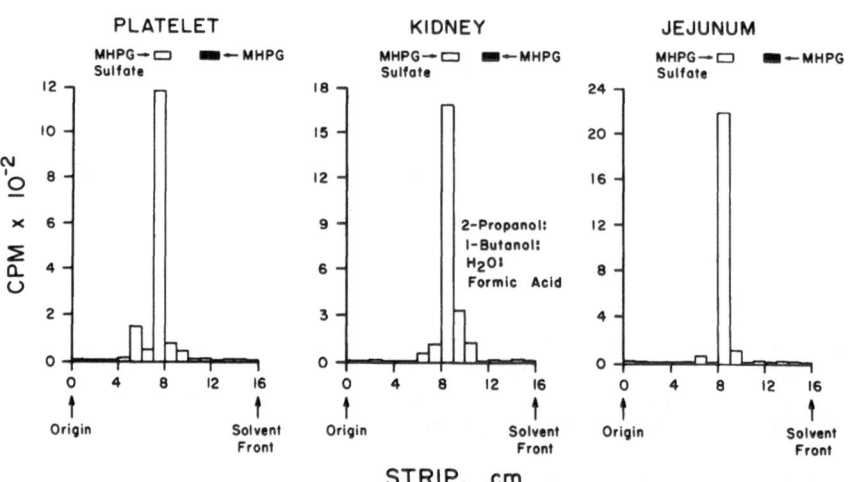

Figure 6. Thin layer chromatography of the product of the PST reaction. The open bar represents the migration of authentic MHPG sulfate and the closed bar represents the migration of MHPG. Counts per minute (CPM) represent the net radioactivity after values for blank samples were subtracted. (Reproduced with the permission of Elsevier/North Holland Publishing Company.) (Anderson and Weinshilboum, 1980)

of 0.0625% BSA. Therefore, this concentration of BSA was used for the dilution of tissue homogenates. These two simple precautions, the addition of dithiothreitol and BSA resulted in, for example, a 4.4 fold increase of the enzyme activity in platelet preparations.

Identification of reaction product by thin layer chromatography. Supernatants from assays of platelet, kidney, and gut homogenates were lyophilized and were used for product identification by thin layer chromatography. 96%, 94%, and 90% of the net radioactivity applied to the thin layer chromatography plates for platelet, kidney, and gut homogenates, respectively, migrated with authentic MHPG sulfate in a solvent system of 2-propanol, 1-butanol, water, and formic acid (60:20:19:1, vol:vol:vol:vol) (Figure 6).

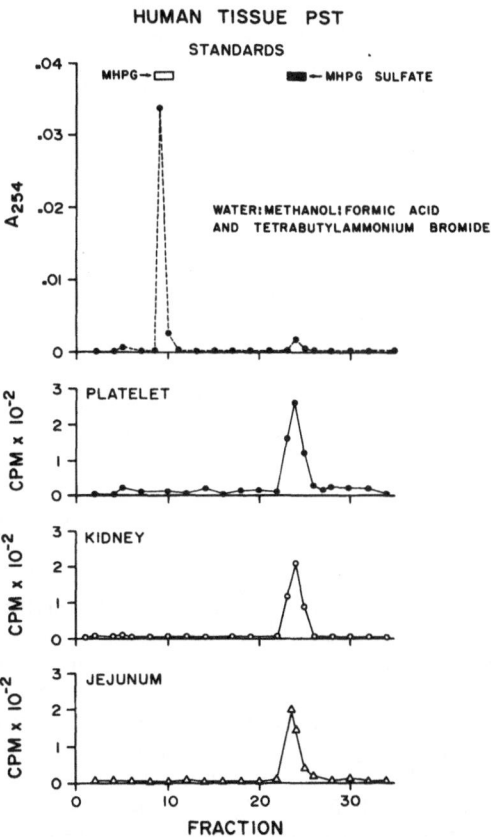

Figure 7. High performance liquid chromatography of the product of the PST reaction. Counts per minute (CPM) represent the net radioactivity eluted after values for blank samples were subtracted. (Reproduced with the permission of Elsevier/North Holland Publishing Company.) (Anderson and Weinshilboum, 1980)

Identification of reaction product by high performance liquid chromatography. Aliquots of supernatants from assays of platelet, kidney, and gut homogenates were mixed with authentic MHPG sulfate and were used for product identification by high performance liquid chromatography. Approximately 100% of the net radioactivity from each active sample eluted from the HPLC column with authentic MHPG sulfate (Figure 7).

Endogenous PST inhibitors. Since potent inhibitors of PST were present in human erythrocyte lysates and renal cortical homogenates, the possible effects of PST inhibitors on the enzyme activity in homogenates of human platelet, renal cortex, and jejunum under these assay conditions were determined. In the first series of experiments, 10 μl aliquots of partially purified human erythrocyte PST were added to platelet homogenates from 10 adult white subjects. The average recovery of the added enzyme activity was 103 + 2.8% (mean + SEM). Recoveries of partially purified PST added to homogenates of renal and gut tissues were 111 + 9.5% and 99 + 1.0% (mean + SEM), respectively. A second series of experiments was performed in which pooled samples of platelet, gut, and kidney homogenates with either high or low levels of enzyme activity were mixed in equal volumes and in proportions of 3:1 and 1:3. The enzyme activities measured in each of the mixtures was virtually identical to the anticipated values based on simple additive contributions by each pooled sample (Anderson and Weinshilboum, 1980). The quantitative recovery of partially purified PST activity from diluted tissue homogenates, and the close agreement of actual and expected activities in the mixing experiments made it extremely unlikely that endogenous PST inhibitors or activators affected the levels of enzyme activity measured in human platelet, kidney, and gut homogenates under these assay conditions.

Platelet PST: Characteristics and Expression of Results

Platelet storage and preparation. The effect of storage while frozen on platelet PST activity was determined. Platelet samples from 10 randomly selected blood donors were prepared and aliquots were either assayed immediately or were stored overnight at -85°C. Only 78 + 12% (mean + SD) of the initial enzyme activity remained in samples that had been frozen overnight. The range of activity remaining in frozen samples varied from 55.1 to 95.8% when compared with the activity in the same samples assayed immediately. Mean platelet PST activity measured immediately after blood was obtained from 50 adult blood donors was 6.6 + 1.6 units per mg platelet protein (mean + SD). All subsequent data are based on the results of assays of homogenates prepared from freshly obtained platelet preparations.

PST is usually considered to be a "soluble" or cytoplasmic enzyme (Dodgson, 1977). To determine whether platelet PST was

soluble, platelet homogenates from 10 randomly selected subjects
were prepared and aliquots were centrifuged at 100,000 x g for 1
hour. The high-speed supernatants retained 101 + 5.7% (mean +
SD) of the enzyme activity present in uncentrifuged homogenates.
Platelet PST also appeared to be a soluble enzyme. Since there
might be individual variations in the protein concentrations of
platelet components removed by centrifugation, expression of
assay results in terms of mg of soluble platelet protein instead
of mg of total homogenate protein appeared preferable. When 47
of the 50 platelet preparations described above were centrifuged
at 100,000 x g for 1 hour, the average protein concentration in
the final supernatants was only 61% of that in uncentrifuged
samples. The mean platelet PST activity expressed per mg soluble
platelet protein was 10.8 + 2.4 units per mg protein (mean + SD)
for these 47 samples.

Coefficient of variation and expression of assay results.
Expression of platelet PST activity on the basis of mg of
platelet protein might introduce variance into the assay results
unrelated to variation in PST activity since platelet pellets
might be contaminated with plasma proteins. Therefore, multiple
PST assays were performed with platelets isolated from the blood
of a single individual and the results were expressed both per
10^8 platelets and per mg soluble platelet protein. One unit of
blood from a 40 year old male donor was collected in acid-
citrate-dextrose, and 10 separate 8 ml aliquots of blood were
placed in plastic centrifuge tubes. Platelet-rich plasma was
prepared from each 8 ml sample, and PST activity was measured.
The results for each separate aliquot were expressed both as
units per mg soluble platelet protein and as units per 10^8
platelets. The coefficient of variation for the assay was 5.8%
when the enzyme activity was expressed per 10^8 platelets, and
was 9.4% when expressed per mg soluble platelet protein. Because
the coefficient of variation was lower for the "platelet count"
method, this method for the expression of the results was used in
subsequent experiments. These results were similar to those of
Jackman et al. (1979) in which they reported that it was
preferable to express monoamine oxidase activity on the basis of
platelet number rather than on the basis of platelet protein.

Platelet PST activity in a randomly selected population.
The previous experiments showed that the variance of the assay
was decreased when the results were expressed in terms of the
number of platelets and that it was necessary to assay platelet
PST activity immediately. It was important to establish a
"normal" range of values for PST activity in platelet samples
assayed in this fashion. PST activity was measured in platelets
from 102 randomly selected adult blood donors. The mean activity
was 1.2 + 0.4 units per 10^8 platelets (mean + SD), with a range
from 0.2 to 2.9. There was a 5 fold intersubject variation of

enzyme activity within + 2 standard deviations. The mean
activity for female subjects was 1.2 + 0.5 units per 10^8
platelets (mean + SD, n = 50), and the mean activity for male
subjects was 1.1 + 0.4 units per 10^8 platelets (mean + SD, n =
52).

   Variation in platelet PST activity with time. Platelet PST
activities in blood samples from 5 individuals were measured
three separate times during a two week period to determine
whether the enzyme activity varied with time. All subjects were
laboratory personnel who were not taking medication and were
neither acutely nor chronically ill. All blood samples were
obtained from fasting subjects between 8-10 AM. The average PST
activity in the initial assays of these 5 subjects was 1.1 + 0.3
units per 10^8 platelets (mean + SD) with a range from 0.8 to
1.7. When the enzyme activities measured at one and two weeks
were expressed as a percentage of the initial PST activities,
values for 2 subjects remained relatively stable, those for 2
subjects increased by approximately 30%, and the enzyme activity

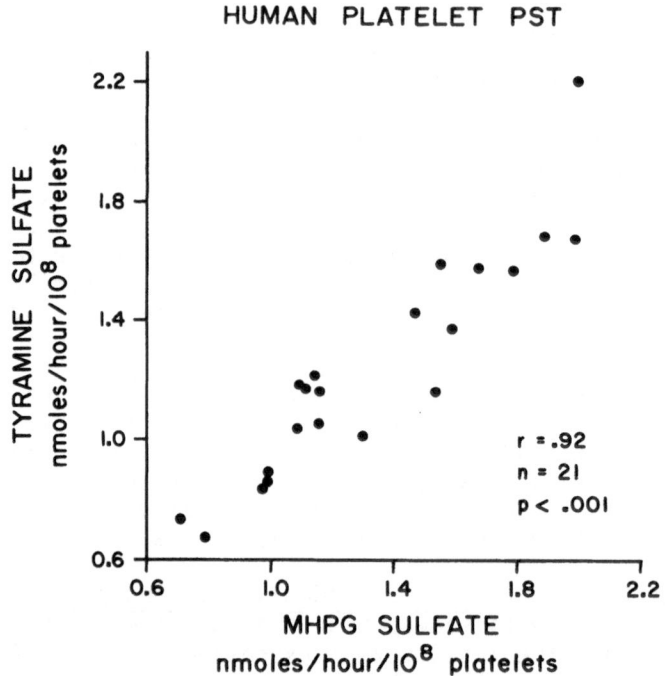

Figure 8. Correlation between sulfate conjugation of MHPG and
tyramine by individual platelet homogenates. Results are
expressed as nmoles of sulfated product per 10^8 platelets per
hour.

in 1 subject decreased by more than 50%. These results suggested
that values of platelet PST activity in individual subjects may
vary widely within a few weeks time. The factors responsible for
this variation remain to be determined.

Platelet PST: Substrate and Tissue Correlations

Correlation between sulfate conjugation of MHPG and other
substrates. Human platelet PST catalyzes the sulfate conjugation of a variety of phenolic substrates (Hart et al., 1979). It
was important to determine whether individual variations in the
platelet enzyme activity determined with MHPG as substrate
reflected variations in the relative degree of sulfate
conjugation of other substrates. Therefore, experiments were
performed to compare the relative platelet enzyme activity
measured with MHPG with the relative enzyme activity measured
with tyramine, dopamine, 5-hydroxytryptamine, acetaminophen, and
alpha-methyldopa as substrates. The rationale underlying these
experiments was to take advantage of the natural variation of
platelet PST activity among individual subjects to determine

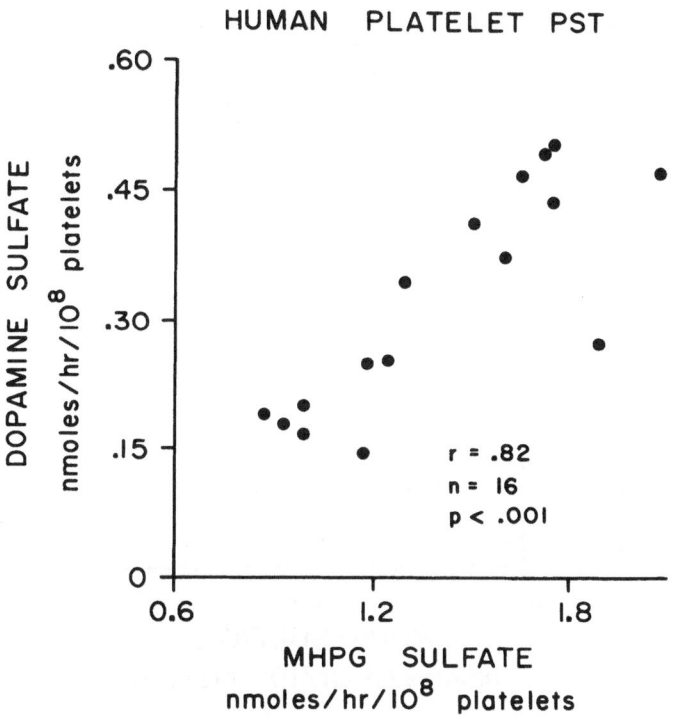

Figure 9. Correlation between sulfate conjugation of MHPG and
dopamine by individual platelet homogenates. Results are
expressed as nmoles of sulfated product per 10^8 platelets per
hour.

24 whether the rank order of the ability of the enzyme in the
platelets of a series of individuals to catalyze the sulfate
conjugation of one substrate were the same for other substrates.
If this were the case, it would support but not prove the
hypothesis that the same enzyme was catalyzing the sulfate
conjugation of each substrate.

Aliquots of platelet homogenates from separate groups of
randomly selected blood donors were assayed with MHPG and with
one of the alternative substrates. Final concentrations for the
substrates were 1.0 mM for tyramine, dopamine, 5-hydroxytrypt-
amine and alpha-methyldopa and 0.5 mM for acetaminophen. Pargyline, a monoamine oxidase inhibitor, at a concentration of 1.0 mM
was included in the reaction mixture when tyramine, dopamine, and
5-hydroxytryptamine were used as substrates. There were significant correlations between the formation of MHPG sulfate and the
formation of sulfated products for each of the other substrates.
The correlation coefficients were: tyramine, r=0.92,n=21 (Figure
8); dopamine, r=0.82, n=16 (Figure 9); 5-hydroxytryptamine,
r=0.94, n=20 (Figure 10); acetaminophen, r=0.77, n=17; and
alpha-methyldopa, r=0.77, n=17 (P < .001 for each). It should be
emphasized that in these experiments neither optimal reaction
conditions nor the identity of the reaction products were
established for substrates other than MHPG. In subsequent

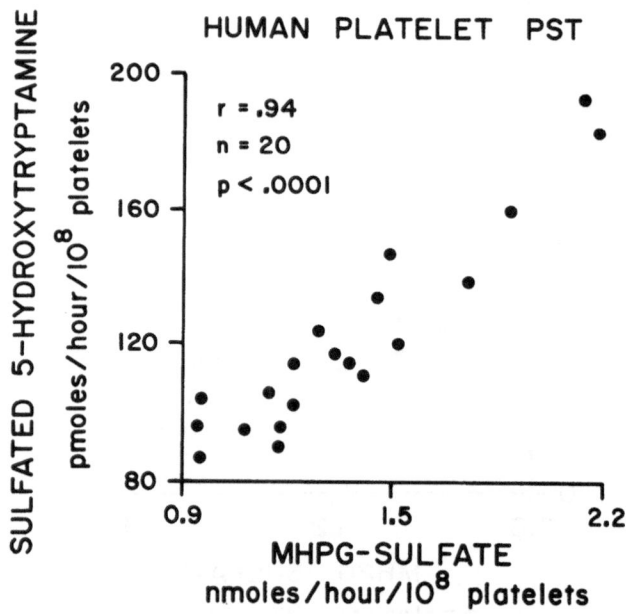

Figure 10. Correlation between sulfate conjugation of MHPG and
5-hydroxytryptamine by individual platelet homogenates. Results
are expressed as nmoles or pmoles of sulfated product per 10^8
platelets per hour.

experiments optimal conditions for the assay of one of the other substrates, acetaminophen, and positive identification of the reaction product by use of high performance liquid chromatography have been accomplished. The correlation experiment was repeated utilizing optimal assay conditions for acetaminophen and the correlation coefficient was still highly significant (r=0.795, n=41, P < .001; Reiter and Weinshilboum, unpublished observation). These results suggested that relative platelet PST activity measured with MHPG reflected the relative ability of platelet homogenates to catalyze the sulfate conjugation of a variety of substrates. The results also suggested that the same enzyme might catalyze the reaction for each substrate.

Correlation between human platelet, kidney and gut PST activities. Relative platelet PST and renal cortical enzyme activities were compared in 20 patients to determine whether the level of platelet enzyme activity reflected the level of PST activity in the kidney. All renal tissues were obtained from patients who underwent clinically indicated nephrectomies for the removal of renal tumors. It required over 8 months to collect the 20 samples of renal tissue even though this study was performed in a very busy surgical center. At the time that the study was initiated, platelet enzyme activity was being expressed per mg of soluble platelet protein. Therefore, all results for this experiment were expressed in that fashion. The mean platelet PST activity per mg soluble platelet protein (9.5 + 4.8 units, mean + SD) was not significantly different from the mean activity of the 47 samples described earlier (10.8 + 2.4 units, mean + SD, P < 0.2). The mean renal cortical PST activity was 1.8 + 0.9 units per mg protein (mean + SD), with a range from 0.4 to 3.7. There was a significant positive correlation between platelet and kidney PST activities (r = 0.54, n = 20, P < 0.02). These results suggested that platelet PST activity might reflect the relative enzyme activity present in the human renal cortex.

Experiments similar to those carried out for the kidney were performed to study the relationship between relative platelet and gut PST activities. Eight human jejunal biopsy samples were obtained over a 5 month period. One sample showed focal villous atrophy, one had increased lamina propria cells with a single crypt abcess and the remainder were histologically normal. The mean PST activity in the gut samples was 81.1 + 37.5 units per mg protein (mean + SD) with a range from 14.5 to 130.4. The correlation coefficient between gut and platelet PST activities was 0.67. Because of the small number of samples studied, this value was not statistically significant (P < 0.1). These tissue samples were not all histologically "normal". The highest PST activity was found in the sample with a single crypt abcess histologically and the lowest was in the sample with focal villous atrophy. Obviously, these results must be interpreted with caution.

PST catalyzes the sulfate conjugation of a variety of monoamine neurotransmitters and phenolic drugs (Richter, 1940; Axelrod et al., 1959; Levy et al., 1975; Saavedra et al., 1975; Alam et al., 1977). Little is known of the possible relationship of individual variations in human PST activity to variations in sulfate conjugation. The results of these experiments represent a significant step toward testing the hypothesis that individual variations in this important enzyme activity might represent one factor in individual variations in the ability to catalyze the sulfate conjugation of both neurotransmitters and drugs. PST activity has been detected in two easily obtained human tissues, the erythrocyte and the platelet (Anderson and Weinshilboum, 1979;1980). The human erythrocyte enzyme has been partially purified and its biochemical characteristics described. The erythrocyte studies have also focused attention on tissue inhibitors of PST (Anderson and Weinshilboum, 1979). Unless assays for the measurement of PST in tissue homogenates have been shown to negate the effects of the inhibitors, inhibitors may contribute the major source of variance in the enzyme activity measured. Optimal assay conditions have been determined for the human platelet, small intestine, and kidney, and the reaction product has been identified by both thin layer and high perform- ance chromatography (Anderson and Weinshilboum, 1980). The development of these sensitive and accurate assays represented a necessary step in the study of individual variations in this enzyme activity in man.

The next steps required determination of optimal procedures for platelet preparation and for the expression of assay results. It also had to be determined whether variations in platelet PST activity measured with MHPG as substrate reflected variations in the sulfate conjugation of other substrates and whether relative platelet PST activity reflected the relative enzyme activity in other tissues. The results indicated that platelet PST activity should be measured immediately and that assay variance was minimized if the activity was expressed per platelet rather than per mg soluble platelet protein. Significant correlations between the ability of human platelet homogenates to catalyze the sulfate conjugation of MHPG and that of a variety of other sub- strates were also shown (Anderson et al., 1980;1981). A signifi- cant correlation between relative platelet PST activity and renal cortical activity was found. A similar trend was present for human intestinal mucosal -- although the results were not statistically significant. These results will make it possible to test the hypothesis that individual variations in platelet PST activity might reflect functional variations in the metabolism of monoamine neurotransmitters and drugs. Unfortunately, changes in platelet PST activity in individual subjects with time may complicate potential clinical applications of measurement of PST platelet activity. Obviously, increased understanding of both the biochemistry and regulation of PST in platelets and other tissues will be required if measurements of the platelet enzyme are to contribute to our understanding of individual variations in drug and neurotransmitter metabolism.

ACKNOWLEDGEMENTS

We thank Luanne Wussow, Joel Dunnette, and our many collaborators for their assistance with these studies. Supported in part by NIH grants NS 11014, HL 17487, and HL 07269. Dr. Weinshilboum is an Established Investigator of the American Heart Association.

REFERENCES

Alam, S.N., Roberts, R.J. and Fischer, L.J. (1977). Age-related differences in salicylamide and acetaminophen conjugation in man. J. Pediatrics, 90, 130-135.

Anderson, R.J. and Weinshilboum, R.M. (1979). Phenol-sulphotransferase: Enzyme activity and endogenous inhibitors in human erythrocyte. J. Lab. clin. Med., 94, 158-171, 1979.

Anderson, R.J. and Weinshilboum, R.M. (1980). Phenolsulphotransferase in human tissue: Radiochemical enzymatic assay and biochemical properties. Clin. chim. Acta, 103, 79-90.

Anderson, R., Weinshilboum, R., Phillips, S. and Broughton, D. (1980). Human platelet phenolsulphotransferase (PST) activity: Correlation with PST activity in the kidney and gut. The Pharmacologist, 22, 301.

Anderson, R.J., Weinshilboum, R.M., Phillips, S.F. and Broughton, D.D. (1981). Human platelet phenol sulphotransferase: Assay procedure, substrate and tissue correlations. Clin. chim. Acta, in press.

Axelrod, J., Kopin, I.J. and Mann, J.D. (1959). 3-Methoxy-4-hydroxyphenylglycol sulfate, a new metabolite of epinephrine and norepinephrine. Biochim. biophys. Acta, 36, 576-577.

Dodgson, K.S. (1977). Conjugation with sulfate. In Drug Metabolism from Microbe to Man, (eds. D.V. Parke and R.L. Smith), Francis Ltd, London, pp. 91-104.

Eisenthal, R. and Cornish-Bowden, A. (1974). The direct linear plot. Biochem. J., 139, 715-720.

Foldes, A. and Meek, J.L. (1973). Rat brain phenolsulfo-transferase -- partial purification and some properties. Biochim. biophys. Acta, 327, 365-374.

Hart, R.F., Renskers, K.J., Nelson, E.B. and Roth, J.A. (1979). Localization and characterization of phenol sulfo-transferase in human platelets. Life Sci., 24, 125-130.

28   Jackman, H., Arora, R. and Meltzer, H.Y. (1979). Comparison of platelet count and platelet protein methods for determination of platelet MAO activity. Clin. chim. Acta, 96, 15-23.

Levy, G., Khanna, N.N., Soda, D.M., Tsuzuki, O. and Stern, L. (1975). Pharmacokinetics of acetaminophen in the human neonate: Formation of acetaminophen glucuronide and sulfate in relation to plasma bilirubin concentration and D-glucaric acid excretion. Pediatrics, 55, 818-825.

Lowry, O.H., Rosebrough, N.J., Farr, A.L. and Randall, R.J. (1951). Protein measurement with the Folin phenol reagent. J. biol. Chem., 193, 265-275.

Reilly, D.K., Rivera-Calimlim, L. and Van Dyke, D. (1980). Catechol-O-methyltransferase activity: A determinant of levodopa response. Clin. Pharmac. Ther., 28, 278-286.

Richter, D. (1940). The inactivation of adrenaline in vivo in man. J. Physiol., 98, 361-374.

Saavedra, J.A., Reid, J.L., Jordon, W., Rawlins, M.D. and Dollery, C.T. (1975). Plasma concentration of alpha-methyldopa and sulphate conjugate after oral administration of methyldopa and intravenous administration of methyldopa and methyldopa hydrochloride ethyl ester. Eur. J. clin. Pharmac., 8, 381-386.

Sladek, S. and Weinshilboum, R. (1980). Human lymphocyte catechol-O-methyltransferase (COMT). The Pharmacologist, 22, 237.

Weinshilboum, R.M. (1978). Human erythrocyte catechol-O-methyltransferase: Correlation with lung and kidney activity. Life Sci., 22, 625-630.

Weinshilboum, R.M. and Raymond, F.A. (1977). Inheritance of low erythrocyte catechol-O-methyltransferase activity in man. Am. J. hum. Genet., 29, 125-135.

Wilkinson, G.N. (1961). Statistical estimations in enzyme kinetics. Biochem. J., 80, 324-332.

Sulfation of MHPG and some other Substrates by Human Platelet Phenolsulfotransferase, PST

E J M Pennings, J L Van Brussel, J G Zanen and G M J Van Kempen

Biochemical Laboratory, Psychiatric Hospital Endegeest, POB 1250
2340 BG Oegstgeest, The Netherlands

In mental health research every step in the monoamine metabolism may be of importance. In man the study of the enzymes concerned is hampered because human tissue is not readily available. There is no valid reason to assume identical properties of human and animal enzymes, nor does an animal model exist for mental health research. Therefore, human enzymes derived from readily accessible sites are urgently needed. Phenolsulfotransferase (PST; EC 2.8.2.1), the enzyme responsible for the sulfation of i.a. catechol compounds, has been found in many animal tissues including brain (Eccleston and Ritchie, 1973; Foldes and Meek, 1973, 1974; Jenner and Rose, 1973; Meek and Neff, 1973; Van Kempen *et al.*, 1975; Wong, 1975).

For the reasons mentioned above the recent demonstration of PST in human blood cells (Anderson and Weinshilboum, 1978, 1979, 1980; Hart *et al.*, 1979) is important as it makes a PST of human origin readily available for *in vitro* studies. This is even more important as there is now some evidence that human platelet PST may serve as a usable model for the study of the enzyme in the human brain (Renskers *et al.*, 1980).

Other enzymes involved in the monoamine metabolism e.g. monoamine oxidase (MAO; EC 1.4.3.4) and catechol-O-methyltransferase (COMT; EC 2.1.1.6) have been studied extensively in human blood cells, with a number of physiological and synthetic substrates. Their localization, kinetic properties, activities in states of health and disease have been reported in numerous papers. There is no clear reason why PST is not of equal importance.

We now present preliminary results which describe some properties of PST from human platelets. We have characterized the enzyme with various sulfate acceptors and we have studied the distribution of enzyme activity in human blood, using density gradients prepared of Percoll. The major part of the PST resides in

29

platelets but 10-20% of the activity is found in the erythrocytes region.

METHODS

Preparation of Adenosine 3'-phosphate 5'-phosphosulphate (PAPS)

PAPS was prepared and purified as described previously (Pennings et al., 1978). The final product was analyzed by high-performance liquid chromatography (Pennings and Van Kempen, 1979). Based on the absorbance at 260 nm it had a purity of at least 99%. PAPS is stable for at least 1 month when stored dry at -20°C, but it is gradually hydrolyzed in aqueous solution, even at this low temperature. The main product of hydrolysis, adenosine 3',5'-bisphosphate, is a strong inhibitor of the sulfotransferase reaction (Roy and Trudinger, 1970). We found that a commercially available preparation of PAPS (P-L Biochemicals, Milwaukee, WI, USA) contained considerable amounts of impurities, even when analyzed on the day of receipt. The main contaminant was adenosine 3',5'-bisphosphate. Obviously, such preparations of PAPS ought to be purified before use, especially when kinetic properties of PST are being studied.

Substrate inhibition by the sulphate acceptor appears to be a common feature of PST of whatever origin (Anderson and Weinshilboum, 1980; Banerjee and Roy, 1968; Foldes and Meek, 1973; Pennings et al., 1978; Renskers et al., 1980), but apparent substrate inhibition by PAPS has been reported in only a few cases (Anderson and Weinshilboum, 1979, 1980; Renskers et al., 1980). In our view the latter effect is caused by the degradation products of PAPS, in particular adenosine 3',5'-bisphosphate. This view is supported by our results obtained with commercial PAPS preparation and by those obtained by Sekura et al. (1979) and by Wong and Yeo (1979), who also found contaminants in commercially available PAP^{35}S preparations.

Isolation of Human Platelets

Outdated platelet rich plasma was centrifuged for 16 min at 20°C at 200 g ($r_{av.}$ 10 cm) to remove contaminating red and white cells. The resulting supernatant was divided in 5 ml portions. These were centrifuged for 6 min at 1000 g ($r_{av.}$ 10 cm). The pellets were resuspended in 2 ml of iso-osmotic buffer consisting of 7 vol of 0.154M-NaCl and 3 vol of 0.154M-Tris-HCl pH 7.4, and containing 6mM-Na$_2$EDTA. The centrifugation step at 1000 g was repeated and the pellets were resuspended to a final platelet concentration of approx. 5.10^8/ml in a buffer consisting of 4 vol of 0.154M-NaCl and 1 vol of 0.154M-Tris-HCl pH 7.4. To study the relationship between the rate of the PST reaction and the enzyme concentration. A preparation of approx. 10^9 platelets per ml was used. Platelet preparations were stored at -80°C.

Assay of PST Activity

PST was assayed with 3-methoxy-4-hydroxyphenylethylene glycol (MHPG; final concentration 0.75mM), 4-methylumbelliferone (MU; final concentration 0.05mM) or 3,4-dihydroxybenzoic acid (final concentration 1.0mM) as the sulfate acceptor. The incubation mixture (1.5 ml; final pH 7.4) also contained: 0.075M-Tris-HCl; 24mM-KH$_2$PO$_4$; 0.1mM-PAPS; 3mM-dithiothreitol; 1mM-MgCl$_2$; 6.25 μl 20% Triton X-100; 0.25 ml of the platelet suspension, containing approx. 125.10^6 platelets. Incubation was performed at 37°C for 1 h. The amount of sulfated product and the ratio between 3-0- and 4-0-sulfated dihydroxybenzoic acid was determined as described previously (Van Kempen and Jansen, 1972; Van Kempen *et al.*, 1975; Pennings and Van Kempen, 1980).

Density Gradient Centrifugation of Human Blood

For extended methodology we refer to recent reviews in Separation News (1979, 1980). Percoll, a density gradient medium based on colloidal silica coated with polyvinylpyrrolidone was obtained from Pharmacia Fine Chemicals, Uppsala, Sweden. A stock solution was prepared by diluting it with 1/9 vol of 1.5M-NaCl. Before use, this stock of iso-osmotic Percoll was diluted with 0.15M-NaCl to a final concentration of 62% (v/v) resulting in a density of approximately 1.086 g/ml. The gradient was formed by centrifugation of 8.0 ml of 62% (v/v) Percoll for 20 min at 4°C at 20.000 g (r$_{av.}$ 5.9 cm) in a Spinco Model L ultracentrifuge (Beckman) with a fixed angle type 40 rotor.

Calibration of the gradient was done by measuring the distance between the bands formed from Density Marker Beads (Pharmacia Fine Chemicals, Uppsala, Sweden). They were added to a control tube that was treated identically to the experimental tubes.

Human blood from healthy subjects was collected with Na$_2$EDTA (to a final concentration of 6mM) as the anticoagulant and diluted with an equal volume of 0.15M-NaCl. Of the resulting suspension, 2 ml was layered on the top of the gradient and the tube was centrifuged for 8 min at 4°C and 1250 g (r$_{av.}$ 5.9 cm). Fractions of 0.5 ml were pipetted from the top of the gradient. These were assayed for protein, enzyme activities and blood cells.

PST activity was assayed as described above with the exception that with MU as the substrate the reaction was terminated by the addition of 0.25 ml of 5M-trichloroacetic acid to precipitate Percoll and proteins. With MHPG as the substrate the incubation was performed at the optimal pH of 8.0.

Monoamine oxidase was determined with kynuramine as the substrate (Kraml, 1965; McEntire *et al.* 1979). The reaction was terminated by the addition of 0.5 ml of 1M-trichloroacetic acid.

Thrombocytes were counted in Bürker chambers.

Erythrocytes were quantitated by measurement of hemoglobin in each fraction.

Protein was assayed by the Lowry (1951) method with horse serum albumin as a standard.

RESULTS

Some Properties of PST from Human Platelets

MHPG and MU as the sulfate acceptor. Addition of 1mM-MgCl$_2$ to the incubation mixture caused an insignificant (0-3%) stimulation of enzyme activity.
In the absence of dithiothreitol the sulfotransferase activity in the platelet preparations began to decrease after 2 to 3 weeks of storage at -80°C. The activity could be restored by addition of 3mM-dithiothreitol to the incubation mixtures. Repeated freezing and thawing caused an additional decrease of activity, which could also be restored by the addition of dithiothreitol. Fresh platelet preparations did not need dithiothreitol for full activity.
The rate of the sulphotransferase reaction was linear with time for at least 1 h and also with the thrombocyte concentration from 35.10^6/ml up to 540.10^6/ml incubation mixture. This indicates that high-molecular-weight inhibitors are absent from human platelets. The pH optimum of the sulfotransferase reaction at 37°C in 0.075M-Tris-HCl buffer was near 8.0 when determined with MHPG and near 7.4 when determined with MU as the substrate (Figure 1).

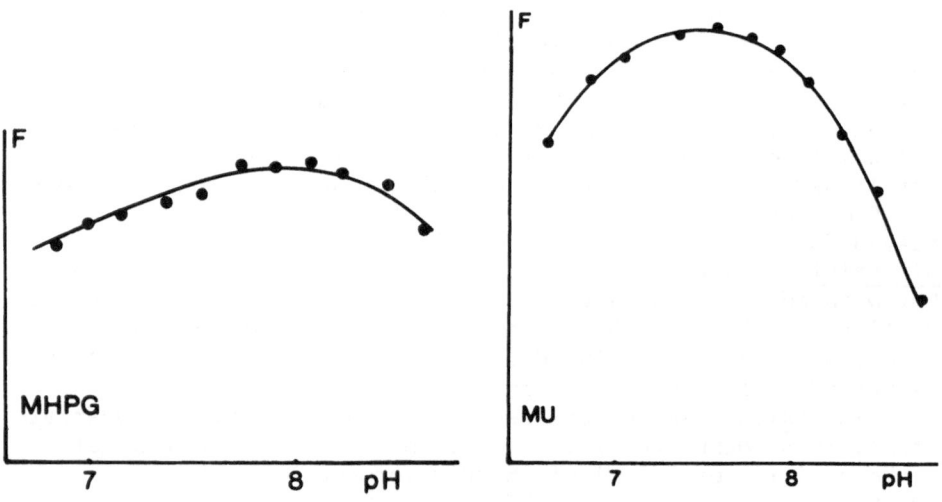

Figure 1.   pH Optima for PST with MHPG or MU as
the substrate

We observed a 3-fold variation of enzyme activity in platelet preparations from different subjects. The mean ($\pm$ S.D.) activity was $(31 \pm 9).10^{-9}$nmol MHPG-sulfate formed/thrombocyte/h (range $17.10^{-9} - 46.10^{-9}$; n=12) and $(29 \pm 10).10^{-9}$nmol MU-sulfate formed/thrombocyte/h (range $12.10^{-9} - 39.10^{-9}$; n=11). In one experiment the ratio of MHPG-sulfation to MU-sulfation was determined in platelet preparations from 9 different persons. The ratio varied from 0.9 to 2.2.

3,4-Dihydroxybenzoic acid as the sulfate acceptor. With this substrate the ratio of 3-0 to 4-0-sulfation was determined. The ratio found was approximately 5. The total activity was approximately $4.7 . 10^{-9}$nmol sulfate ester formed/thrombocyte/h.

Kinetic studies of PST from Human Platelets

Substrate inhibition by MHPG and MU was apparent in concentrations over 0.75mM and 0.05mM respectively. We never observed inhibition by PAPS in concentrations up to 2.5mM.

The type of substrate inhibition was studied from replots of slopes and intercepts of double reciprocal plots (Cleland, 1970). With either acceptor the type of inhibition was uncompetitive. This indicates an ordered addition of substrates to the enzyme, PAPS being the first substrate to add. With this assumption we calculated the various kinetic constants for PAPS, MHPG and MU.

MHPG as the substrate. The mean values (n=2) of K_sPAPS, K_mPAPS and K_mMHPG were 8.5 μM, 8.5 μM and 0.33mM respectively. The mean maximal velocity was 50.10^{-9}nmol MHPG-sulfate formed/thrombocyte/h.

MU as the substrate. The mean values (n=3) of K_sPAPS, K_mPAPS and K_mMU were 17 μM, 12 μM and 12 μM respectively. The mean maximal velocity was 60.10^{-9}nmol MU-sulfate formed/thrombocyte/h.

Arylsulfatase in thrombocytes. As Metcalfe et al. (1979) recently demonstrated the presence of arylsulfatases in human platelets, we did some preliminary experiments on this enzyme. We could quantitatively demonstrate that MU-sulfate at pH 5.6 was hydrolyzed by thrombocytes. Our results do, however, not yet permit a quantitative interpretation. In the PST experiments, at a pH between 7 and 8 and in the presence of fosfate as sulfatase inhibitor, we did not find any hydrolysis of the sulfated compounds formed.

Density Gradient Centrifugation of Human Blood

Whole human blood can be separated into its cell components on density gradients of Percoll (Pertoft et al., 1968; Segal et al., 1980) as can be seen from Figures 2 and 6. MAO can be used as a marker for thrombocytes as it is apparently absent from blood plasma and erythrocytes (Tipton, 1979). The information obtained from these experiments did, however, not allow us to make a clearcut decision on the presence or absence of PST in erythrocytes. Therefore we performed some experiments with puri-

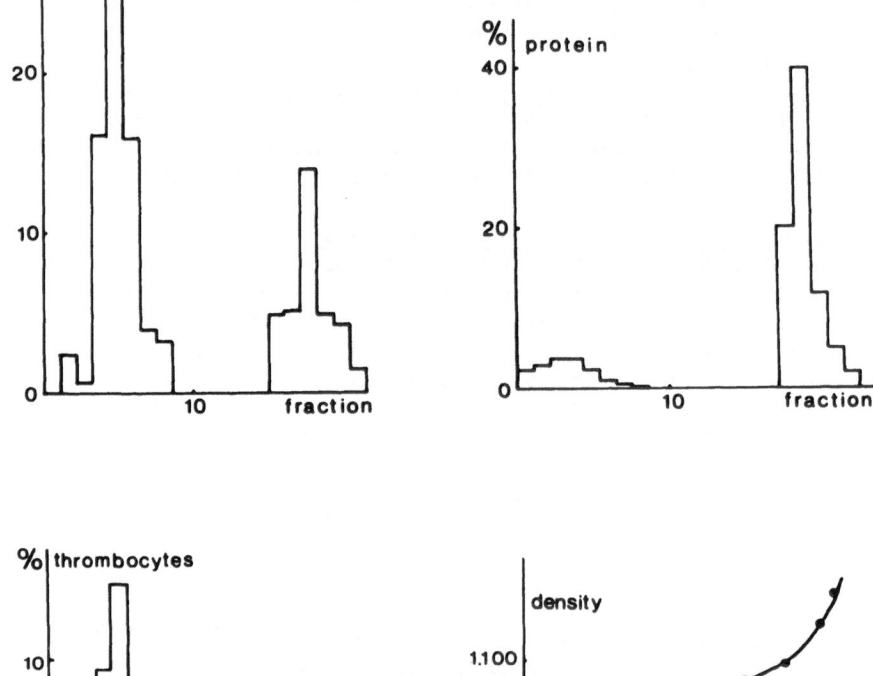

Figure 2.   Distribution pattern of PST activity,
thrombocytes, protein and density after
centrifugation of whole blood on a Percoll gradient.

fied preparations of thrombocytes and erythrocytes.
A purified thrombocyte preparation was fractionated on the Percoll gradient and in all the 20 fractions thrombocytes, protein,
MAO and PST - with the two substrates - were measured. All PST
activity was found in the thrombocytes containing fractions and
this activity apparently parallels that of MAO, as can be seen
in Figure 3.
The results with erythrocytes are shown in Figure 4. The washed
erythrocytes did not contain measurable PST activity towards

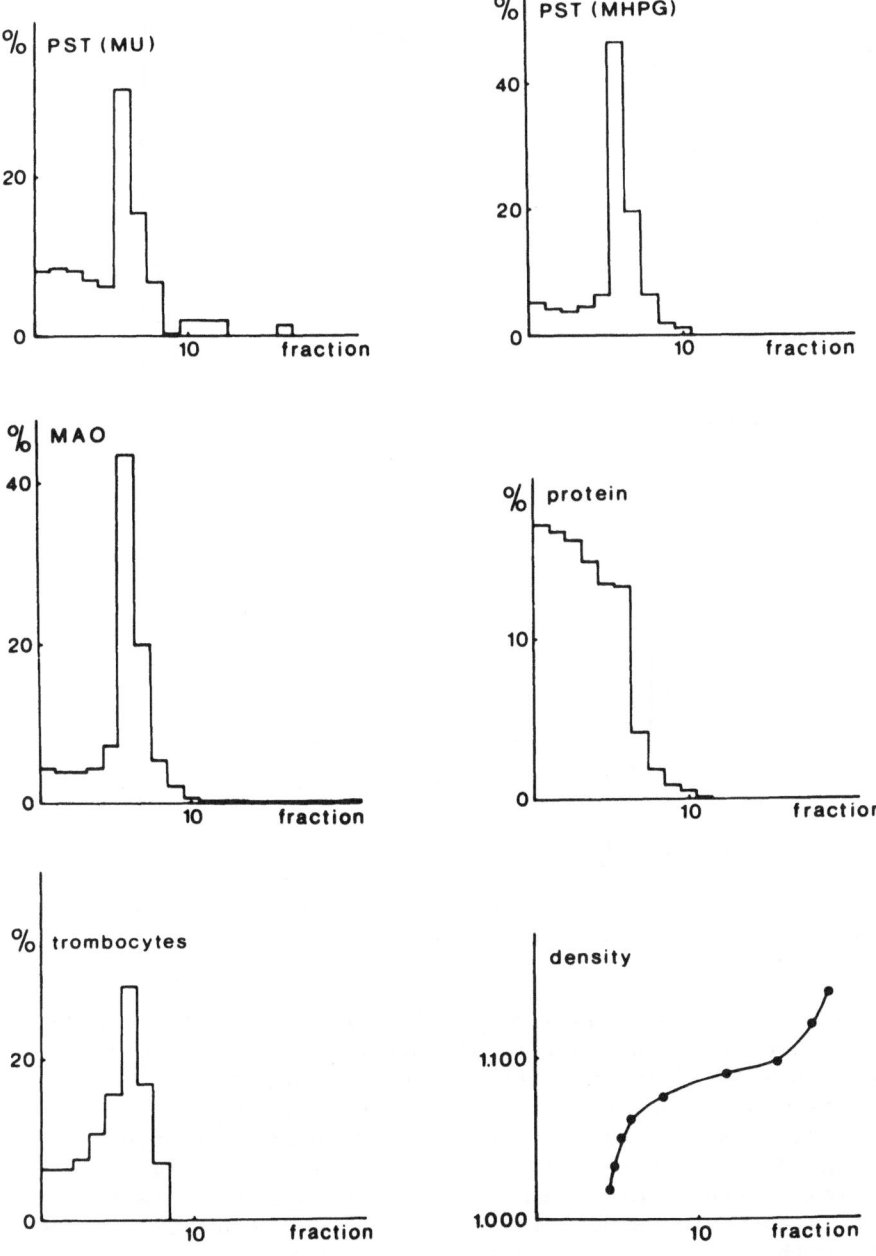

Figure 3. Distribution pattern of enzyme activities, thrombocytes, protein and density after centrifugation of a purified thrombocyte preparation on a Percoll gradient.

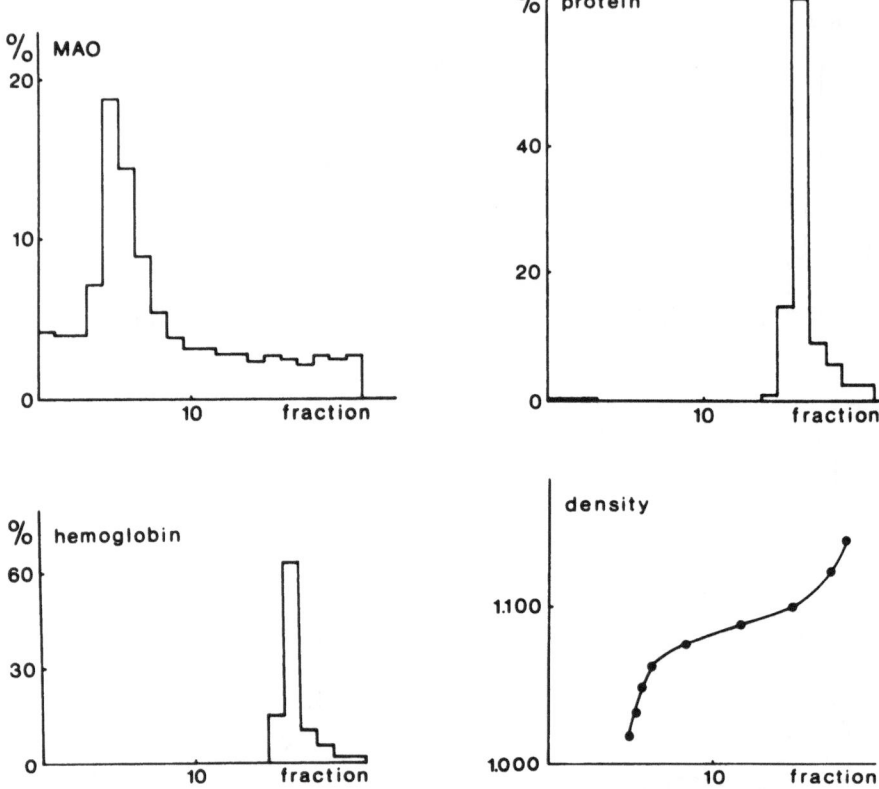

Figure 4.   Distribution of MAO, protein, hemoglobin and
             density after centrifugation of a purified
             erythrocyte preparation on a Percoll gradient.
             In this preparation no PST activity could be
             demonstrated.

either substrate. The observed very low MAO activity is localized in those fractions that in all other experiments contain
the thrombocytes. This MAO activity must be attributed to a
small contamination of the erythrocytes with thrombocytes.
From Figures 3 and 4 the activities in the different preparations can not be compared because the figures show only relative
activities. We therefore present the activities or contents, expressed in absolute values, in Figure 5. The MAO activity found
in the erythrocyte preparation appears to be only very small in
relation to that in the thrombocyte preparation, whereas for
protein the opposite is true.
Segal *et al.* (1980) described that defibrination of blood samples by agitation in the presence of glass beads resulted in the
disappearance of platelets and that the residual cells could be

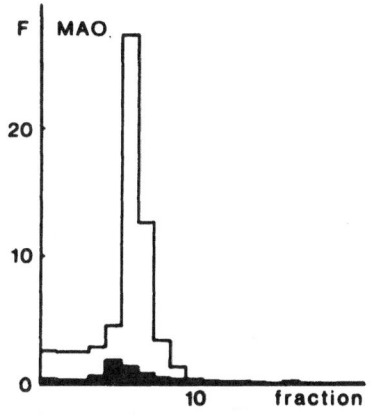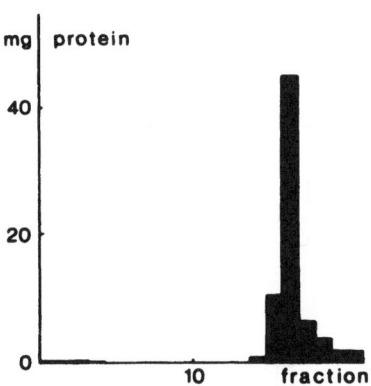

Figure 5. Distribution of MAO and protein - in units
of activity or milligrams - after centrifugation of erythrocytes (black) or thrombocytes (white) on a Percoll gradient.

succesfully separated on Percoll gradients. We tried this method
to clarify the cellular localization of PST in human blood cells.
The results are shown in Figure 6. In this experiment the whole
blood sample used was not the same as that used for the experiment given in Figure 2.

DISCUSSION

Recent studies have described the presence of PST in human blood
(Anderson and Weinshilboum, 1978, 1979, 1980; Hart *et al.*, 1979).
Our report confirms and extends these results. It is now clear
that blood can be an important site of sulfoconjugation. This
agrees with the finding that catecholamines in human blood plasma are for their major part conjugated with sulfate (Johnson *et
al.*, 1980).
From the results presented here we do not yet obtain a clear pic-
ture of the sulfoconjugation system in human blood cells. A purification of enzyme(s) seems necessary as some results indicate
the presence of more than one sulfotransferase. We found for example an interindividual variation in the ratio between the sulfotransferase activity towards MU and MHPG, which can not be explained by the existence of varying amounts of endogenous inhibitors. The presence of inhibitors in thrombocytes is also unlikely as we always have measured recoveries of 100 ± 20% after
gradient centrifugation and as we have confirmed the linear relationship between the rate of the sulfotransferase reaction and
the concentration of thrombocytes, as described already by Anderson and Weinshilboum (1980).

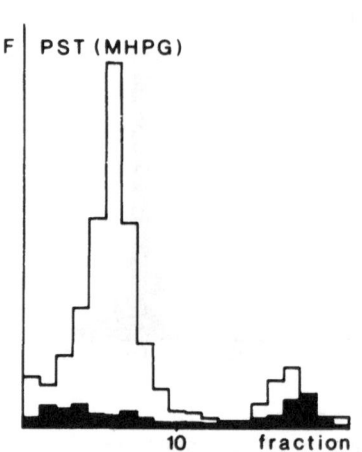

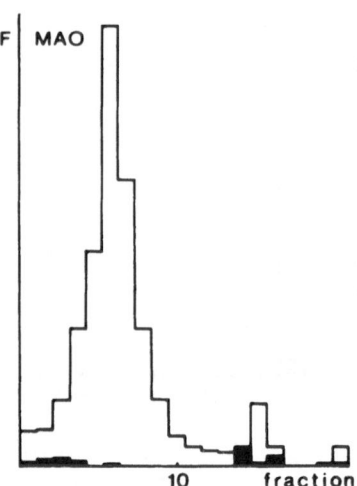

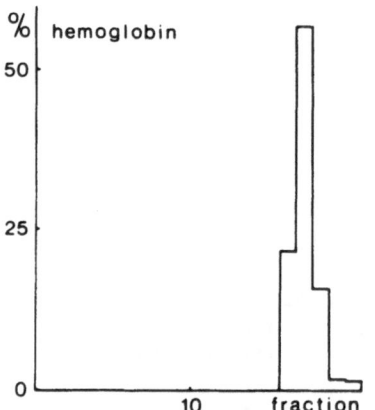

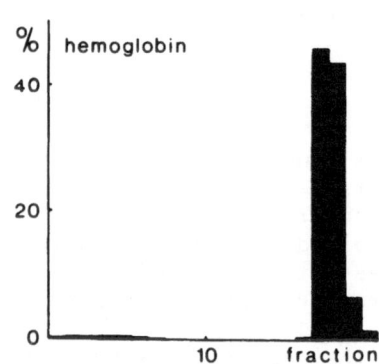

Figure 6. Distribution of PST, MAO and hemoglobin in whole blood before (white) or after (black) defibrination.

A sequential ordered Bi Bi mechanism for the sulfotransferase reaction is indicated by the uncompetitive substrate inhibition that occurs with both sulfate acceptors. Therefore we expected a K_S value for PAPS independent of the acceptor used. This was, however, not confirmed experimentally. More experiments are needed to asses this finding and to answer the question whether a different mechanism applies to platelet PST or platelets contain more than one sulfotransferase.

The pH optimum of the sulfotransferase reaction depends on the acceptor used. This agrees with our previous findings on the rat brain enzyme (Pennings *et al.*, 1978). But we have no simple explanation for the large difference in the K_m values for MHPG and MU.

The K_m value for MHPG and the value of the pH optimum with MHPG as the acceptor both differ from those given by Anderson and Weinshilboum (1980). This can be explained by the difference in the substrate concentrations used. We emphasize that saturating concentrations of pure PAPS must be used. Besides, the concentration of the acceptor substrate must be carefully chosen to avoid substrate inhibition. As we have shown previously the pH optimum of the sulfotransferase reaction can also be seriously affected by the appearance of substrate inhibition (Pennings *et al.*, 1978).

A common problem in the use of blood cells as the enzyme source is the fact that in most cases a preliminary purification is performed. This may give distorted results because the purification procedure often caused a considerable loss in specific types of cells. According to Murphy *et al.* (1978) some larger platelets would seem to be excluded from the conventional preparative procedure for platelet-rich plasma. This may implicate the loss of subpopulations of cells differing in e.g. density, size or age. We therefore have chosen for a method that uses whole blood and that allows us to calculate recoveries in the gradient relative to the activity in whole blood and to that in a mixed gradient. From our results we conclude that most, and probably all, PST as measured with MU and MHPG as the substrates, is localized in the thrombocytes and parallels in this respect MAO. About the origin of the small proportion of the activity found in the erythrocyte region we hesitate to decide. In a purified thrombocyte preparation we did not observe PST or MAO activity in the erythrocyte region. In a purified erythrocyte preparation we found no PST activity at all, and only a very small MAO activity in the area of the thrombocytes. After defibrination, almost all PST and MAO activities have disappeared. In our view only a special type of thrombocytes with unexpected sedimentation behavior (see also Murphy *et al.*, 1978) or a subtype of erythrocytes, maybe the very young, are involved. More likely, a small part of the thrombocytes sediments together with the erythrocytes because they are clustered or bound to erythrocytes.

Thrombocytes contain arylsulfatase activity (Metcalfe *et al.*, 1979). The presence of arylsulfatase together with PST in blood platelets seems strange. The primary role of arylsulfatase is, however, probably not the hydrolysis of catecholamine compounds. This is indicated by the high K_m values of arylsulfatase towards dopamine 3-0- and 4-0-sulfate (Jenner and Rose, 1978) and by our observations (unpublished results) that MHPG-sulfate is a very poor substrate for human urinary arylsulfatase.

40 The major role of PST in human blood may be the sulfation of catecholamines and their metabolites. This view is supported by e.g. the high plasma levels of sulfated catechols in normals (Johnson *et al.*, 1980) and of dopamine-sulfate in Parkinson patients during levodopa therapy (Tyce *et al.*, 1974).

The question of the physiological substrate is also relevant in view of the growing criticism on the meaning of the excretion of free and/or conjugated MHPG as an index of central noradrenergic activity. The relative importance of thrombocyte PST in MHPG sulfation and knowledge of which part of urinary MHPG-sulfate is sulfated peripherally by platelet PST still remains unclear. In this respect the high Km value found for MHPG was unexpected. In view of the recent conflicting evidence on the excretion of the optically active isomers of MHPG, it may be of interest to investigate the existence of stereoselectivity in the action of PST (Baillie *et al.*, 1980; Blomberg *et al.*, 1980).

Another physiological substrate for platelet PST, tyramine (Hart *et al.*, 1979) may be of interest in the study of vulnerability to depressive illness (Bonham Carter *et al.*, 1980).

There exists a still increasing number of papers on the role, levels, and metabolism of monoamines and their metabolites in the central nervous system. It is also known for a number of years that one of the reactions that can take place in this metabolism is the conjugation with sulfate. It is therefore surprising to see that the literature on this aspect of monoamine metabolism is rather small. One finds papers describing the effect of a certain compound on noradrenergic activity as measured by the formation or excretion of MHPG-sulfate, but without a study of a possible effect of that compound on the enzymes involved in the metabolic steps between noradrenaline and MHPG-sulfate. It is clear that data on the influence of foreign or endogenous compounds on PST are urgently needed.

A possible influence by state of health and circadian rhythm on PST is as far as we know not yet investigated.

If, and it is suggested by the work of Renskers *et al.* (1980), the human platelet may serve as a good model for the transferase activity in human brain, a number of these questions may be answered within the next few years.

REFERENCES

Anderson, R.J. and Weinshilboum, R.M. (1978). Phenolsulphotransferase (PST): presence of enzyme activity and of endogenous inhibitors in the human erythrocyte (RBC). Pharmacologist, 20, 218.

Anderson, R.J. and Weinshilboum, R.M. (1979). Phenolsulphotransferase: enzyme activity and endogenous inhibitors in the human erythrocyte. J. Lab. Clin. Med., 94, 158-171.

Anderson, R.J. and Weinshilboum, R.M. (1980). Phenolsulphotransferase in human tissue: radiochemical enzymatic assay and biochemical properties. Clin. Chim. Acta, 103, 79-90.

Baillie, T.A., Boobis, A.R., Davies, D.S., Frank, H. and Murray, S. (1980). Stereoselective excretion of 3-methoxy-4-hydroxyphenylethylene glycol sulphate in the dog. Brit. J. Pharmacol., 69, 320-321.

Banerjee, R.K. and Roy, A.B. (1968). Kinetic studies of the phenol sulphotransferase reaction. Biochim. biophys. Acta, 151. 573-586.

Blombery, P.A., Kopin, I.J., Gordon, E.K., Markey, S.P. and Ebert, M.H. (1980). Conversion of MHPG to vanillylmandelic acid. Arch. Gen. Psychiatry, 37, 1095-1098.

Bonham Carter, S.M., Reveley, M.A., Sandler, M., Dewhurst, J., Little, B.C., Hayworth, J. and Priest, R.G. (1980). Decreased urinary output of conjugated tyramine is associated with lifetime vulnerability to depressive illness. Psychiatry Res., 3, 13-21.

Cleland, W.W. (1970). Steady state kinetics. In The Enzymes (ed. P.D. Boyer), vol. II, Academic Press, New York.

Eccleston, D. and Ritchie, I.M. (1973). Sulphate ester formation from catecholamine metabolites and pyrogallol in rat brain in vivo. J. Neurochem., 21, 635-646.

Foldes, A. and Meek, J.L. (1973). Rat brain phenolsulfotransferase - partial purification and some properties. Biochim. biophys. Acta, 327, 365-373.

Foldes, A. and Meek, J.L. (1974). Occurrence and localization of brain phenolsulphotransferase. J. Neurochem., 23, 303-307.

Hart, R.F., Renskers, K.J., Nelson, E.B. and Roth, J.A. (1979). Localization and characterization of phenol sulfotransferase in human platelets. Life Sci., 24, 125-130.

Jenner, W.N. and Rose, F.A. (1973). Studies on sulphation of 3,4-dihydroxyphenylethylamine (dopamine) and related compounds by rat tissues. Biochem. J., 135, 109-114.

Jenner, W.N. and Rose, F.A. (1978). Dopamine 3- and 4-0-[35s] sulphates as substrates for arylsulphatases in vitro and their metabolism by the rat in vivo. In Conjugation Reactions in Drug Biotransformation (ed. A. Aitio), Elsevier/Norh Holland Biochemical Press, Amsterdam.

Johnson, G.A., Baker, C.A. and Smith, R.T. (1980). Radioenzymatic assay of sulfate conjugates of catecholamines and dopa in plasma. Life Sci., 26, 1591-1598.

Kraml, M. (1965). A rapid microfluorimetric determination of monoamine oxidase. Biochem. Pharmac., 14, 1683-1685.

Lowry, O.H., Rosebrough, N.J., Farr, A.L. and Randall, R.J. (1951). Protein measurement with the Folin phenol reagent. J. biol. Chem., 193, 265-275.

McEntire, J.E., Buchok, S.J. and Papermaster, B.W. (1979). Determination of platelet monoamine oxidase activity in human platelet-rich plasma. Biochem. Pharmacol., 28, 2345-2347.

42   Meek, J.L. and Neff, N.H. (1973). Biogenic amines and their metabolites as substrates for phenol sulphotransferase (EC 2.8.2.1) of brain and liver. J. Neurochem., 21, 1-9.

Metcalfe, D.D., Corash, L.M. and Kaliner, M. (1979). Human platelet arylsulphatases: identification and capacity to destroy SRS-A. Immunology, 37, 723-729.

Murphy, D.L., Costa, J.L., Shafer, B. and Corash, L. (1978). Monoamine oxidase activity in different gradient fractions of human platelets. Psychopharmacology, 59, 193-197.

Pennings, E.J.M., Vrielink, R. and Van Kempen, G.M.J. (1978). Kinetics and mechanism of the rat brain phenol sulphotransferase reaction. Biochem. J., 173, 299-307.

Pennings, E.J.M. and Van Kempen, G.M.J. (1979). Analysis of 3'-phosphoadenylylsulphate and related compounds by paired-ion high-performance liquid chromatography. J. Chromatogr., 176, 478-479.

Pennings, E.J.M. and Van Kempen, G.M.J. (1980). Studies on the *meta* and *para* 0-sulphation of the catechol compound 3,4-dihydroxybenzoic acid by rat liver sulphotransferase *in vitro*. Biochem. J., 191, 133-138.

Pertoft, H. and Lindahl-Kiessling, K. (1968). Separation of various blood cells in colloidal silica-polyvinylpyrrolidone gradients. Exp. Cell Res., 50, 355-368.

Pharmacia Fine Chemicals (Uppsala, Sweden). Separation News, 1979-1, 1980-3.

Renskers, K.J., Feor, K.D. and Roth, J.A. (1980). Sulfation of dopamine and other biogenic amines by human brain sulfotransferase. J. Neurochem., 34, 1362-1368.

Roy, A.B. and Trudinger, P.A. (1970). The biochemistry of inorganic compounds of sulphur. The University Press, Cambridge.

Segal, A.W., Fortunato, A. and Herd, T. (1980). A rapid single centrifugation step method for the separation of erythrocytes, granulocytes and mononuclear cells on continuous gradients of Percoll. J. immunol. Methods, 32, 209-214.

Sekura, R.D., Marcus, C.J., Lyon, E.S. and Jakoby, W.B. (1979). Assay of sulfotransferases. Anal. Biochem., 95, 82-86.

Tipton, K.F. (1979). Monoamine oxidase. In The Neurobiology of Dopamine (eds. A.S. Horn, J. Korf and B.H.C. Westerink), Academic Press, New York.

Tyce, G.M., Sharpless, N.S. and Muenter, M.D. (1974). Free and conjugated dopamine in plasma during levodopa therapy. Clin. Pharmacol. Therap., 16, 782-788.

Van Kempen, G.M.J. and Jansen, G.S.I.M. (1972). Quantitative determination of phenolsulfotransferase using 4-methylumbelliferone. Anal. Biochem., 46, 438-442.

Van Kempen, G.M.J., Wolters, W.L. and Van Elk, R. (1975). Distribution of 3-methoxy-4-hydroxyphenylethyleneglycol sulphotransferase in brain fractions. J. Neurochem., 24, 825-827.

Wong, K.P. (1975). The biosynthesis of 3-methoxy-4-hydroxyphenyl-   43
    glycol sulphate by liver and brain. J. Neurochem., <u>24</u>, 1059-
    1063.
Wong, K.P. and Yeo, T. (1979). Assay of adenosine 3'-phosphate
    5'-sulphatophosphate in hepatic tissues. Biochem. J., <u>181</u>,
    107-110.

Monoamine Sulphoconjugation by Human Platelets

Giovanni B Picotti, Andrea M Cesura, Maria D Galva,
Paolo Mantegazza, Rolf Kettler and Mosè Da Prada

Institute of Pharmacology, School of Medicine, University of Milan,
20129 Milan, Italy
Pharmaceutical Research Department, F Hoffmann-La Roche & Co,
CH-4002, Basle, Switzerland

INTRODUCTION

In the last two decades blood platelets have been widely used as an easily accessible peripheral tissue for biochemical, pharmacological and clinical studies of the processes of amine uptake, storage and release (for review see Pletscher, 1978; Da Prada et al., 1981). Until recently, research on amine bio-transformation in platelets was almost exclusively confined to studies of oxidative deamination by monoamine-oxidase (MAO) (Sandler and Youdim, 1972; Murphy, 1976; Youdim and Hefez, 1980). Amine conjugation reaction in platelets have received far less attention.

There have been a few reports on catechol-0-methyltransfer-ase (COMT) activity in human and rat platelets (De Luca et al., 1976; Stramentinoli et al., 1978). Although Born and Smith(1970) reported data supporting the contention that human platelets sul-phoconjugate adrenaline (A), only recently phenol sulphotransfer-ase (PST) activity has been detected in the cytosol fraction of human platelets (Hart et al., 1979; Anderson and Weinshilboum, 1980; Da Prada et al., 1980). This finding stimulated systematic investigation of the characteristics of PST activity in platelet homogenates with various substrates, e.g., mono- and dihydroxyl-ated aromatic amines and their deaminated and/or methoxylated me-tabolites.

However, to further clarify the role of the enzyme in phys-iological and pathological conditions, methodological approaches other than the study of the kinetics of the reaction in vitro in solubilized enzyme preparations, should be used. Studies of ami-ne sulphoconjugation in platelets incubated with labelled amines and the estimation of endogenous concentrations of unconjugated and conjugated amines in platelets and plasma should provide information about PST activity <u>in situ</u>, under conditions in which enzyme configuration and milieu are unchanged. These approaches may yield more precise information about the physiological plate-let amine-sulphoconjugation processes and their relation to the <u>in vitro</u> and <u>in vivo</u> dynamics of amine uptake and storage.

KINETICS OF PLATELET PST ACTIVITY WITH DIFFERENT AMINE SUBSTRATES

Several endogenous phenolic amines, including catecholamines (CA), tyramine and 5-hydroxytryptamine (5-HT) are substrates for PST, an enzyme (or a group of isoenzymes) which catalyzes the transfer of sulphate from 3'-phospho-adenosyl-5'-phosphosulphate (PAPS) to amines, their deaminated metabolites and a variety of phenolic drugs (Eccleston and Ritchie, 1973;Foldes and Meek,1974; Mulder and Scholtens, 1977; Anderson and Weinshilboum, 1980; Renskers et al., 1980). Human platelets have high PST activity, which is mainly localized in the soluble fraction of the cyto-plasm (Hart et al., 1979).

The literature on the kinetics of platelet PST gives only scant information on the affinity of the amine substrates for the enzyme. In general, excessive substrate concentrations cause di-rect or end-product inhibition of the conjugation reaction (Hart et al., 1979).

The apparent K_m values, as determined for human platelet PST with several amine substrates at various concentrations in the

Table 1. Kinetic parameters for human platelet phenol sulphotransferase, for different amine substrates

Substrate	Apparent Km ($\times 10^{-6}$M)	Relative activity (pmol/mg protein/h)
3-Methoxy-tyramine	0.3	154.7
5-Hydroxytryptamine	1.0	97.9
Tyramine	1.5	177.6
Normetanephrine	1.6	160.2
Metanephrine	1.6	145.7
Dopamine	1.9	196.3
Adrenaline	3.0	88.2

The experiments were performed in the presence of ^{35}S-3'-phospho-adenosyl-5'-phosphosulphate (2.2 x 10^{-8}M). Enzyme activity was measured at the following substrate concentrations: 3-methoxy-tyramine (1 x 10^{-7}M), others (1 x 10^{-6}M). Values are means of two experiments in duplicate.

presence of 2.2 x10^{-8}M ^{35}S-PAPS, are listed in Table 1. Under these experimental conditions, the apparent affinities of catechol and phenol amines (e.g., dopamine and tyramine) do not markedly differ. Dopamine has a lower K_m than the β -hydroxylated CA, A (Table 1) or noradrenaline(NA,not shown).The CA-methoxy-derivatives, methoxy-tyramine (MT), metanephrine (MN) and normetanephrine (NMN) have consistently lower K_m than their CA precursors, indicating that O-methylation in the meta position increases the affinity of the catechol moiety for the enzyme.

At the substrate concentrations used, dopamine is the most rapidly esterified substrate, followed by tyramine and then by the CA-methoxy-derivatives (Table 1). This indicates that, at least in the case of tyramine and CA-methoxy-derivatives, platelet PST efficiently catalyzes the transfer of sulphate to OH- -groups in the para position. At present, there are no data on the ratio of meta- to para-O-sulphoconjugation of CA in platelets. It should be noted that the average ratio of meta- to para- -O-sulphate ester formation is 4 to 1 for dopamine in the frontal lobe of human brain (Renskers et al., 1980) and that dopamine meta-O-sulphate is the predominant metabolite in rat plasma and urine (Jenner and Rose, 1973).

AMINE SULPHOCONJUGATION BY INTACT PLATELETS IN VITRO

Isolated platelets take up, store and release amines, providing a suitable experimental model for studying amine metabolism in vitro under conditions in which the ultrastructure of the platelet is maintained. With this approach the pattern of enzyme activity, its intracellular environment and the various steps involved in amine sulphoconjugation (including the generation of co-factors) should closely resemble the situation in vivo.

Human platelets incubated in plasma take up and sulphoconjugate several labelled amines, known as PST substrates (Table 2). Only a part of the radioactivity extracted from platelets is accounted for by unconjugated amines, whereas all the accumulated radioactivity is recovered as free amine only after acid hydrolysis a 100°C. Only small amounts of CA are O-methylated by platelet COMT (Picotti et al., 1981b). Minor amounts of the amines taken up may also undergo deamination by MAO. The deaminated metabolites, as it is well known, readily leave the platelets and can be found in the incubation medium (Pletscher, 1978).

As shown in Table 2, when various amines are incubated at equimolar concentrations, they are taken up and sulphoconjugated by platelets to different extents. Both the absolute amount of conjugated amine formed and the ratio of conjugated to free amine vary, in relation to their different affinities for PST and their different degrees of availability to the conjugating enzymes. The extent of amine penetration into the cell and its subcellular compartmentalization, e.g., sequestration in storage granules,

Table 2. Uptake and conjugation of various radioactive
amines in isolated human platelets

Amine	Total uptake	Free	Conjugated	% Conjugated
[3]H-5-Hydroxytryptamine	40.1	32.1	8.0	20
[3]H-Dopamine	9.9	4.0	4.6	47
[3]H-Adrenaline	1.4	0.8	0.5	36
[3]H-Noradrenaline	2.1	1.1	0.9	43
[3]H-Methoxy-tyramine	22.0	1.1	20.9	95
[14]C-Metanephrine	21.6	1.3	20.3	94
[14]C-Normetanephrine	20.4	1.5	18.9	93

Values are expressed in pmol/mg protein. Platelets (0.5 ml of
EDTA-PRP) were incubated for 30 min at 37°C with 5×10^{-8}M radioactive amines. The platelet radioactivity was extracted by lysis
with 0.3 N perchloric acid. Catecholamines were separated by absorption on alumina (Spano and Neff, 1971), 5-hydroxytryptamine
by column chromatography (Pletscher et al., 1967) and catechol-
amine-methoxy-derivatives by thin layer chromatography (Da Prada
and Zürcher, 1976). The procedure was repeated after acid hydrolysis of the perchloric extract at 100°C to estimate total (free
plus conjugated) amines. Mean of two experiments in duplicate.

are important factors in limiting amine availability to the enzyme active sites. For instance, 5-HT and A have similar rates of
sulphoconjugation in platelet homogenates (Table 1). However, 5-HT
is taken up much more efficiently into intact platelets than A
(Born and Smith, 1970; Da Prada et al., 1980) and is even more
efficiently accumulated in storage organelles (Da Prada and
Pletscher, 1969), where it is protected from enzymatic degradation. Therefore a greater absolute amount but a smaller proportion of 5-HT than of A is extragranular and available for sulphoconjugation (Table 2).

The methoxy-derivatives of CA, NMN, MN and MT, which not only are excellent substrates for platelet PST but also have very
low affinity for storage granules (Picotti et al., 1981b), have
the highest ratio of conjugated to free amine (Table 2). For
these amines, as opposed to 5-HT and CA, for which the mechanism
of granular storage is essential for the formation of a plate-
let/plasma gradient (Da Prada and Picotti, 1979; Lorez et al.,
1979; Da Prada et al., 1980), sulphoconjugation is an alternative
to granular storage, reducing the cytoplasmic concentration of
free amine and thus facilitating inward transport across the plasma membrane. Since the transport of methoxy-derivatives of CA
is mainly by passive diffusion, sulphoconjugation should be re-

Table 3. Effects of various drugs on the in vitro uptake and conjugation of ^{14}C-normetanephrine (NMN) in human platelets

Drug	Conc.(M)	Platelet [14]C- NMN		
		Total uptake	Free	Conjugated
None		337	39	298
Ouabain	10^{-4}	339	-	-
Chlorimipramine	10^{-5}	330	-	-
Cocaine	10^{-4}	301	-	-
NaF	10^{-2}	27	-	-
N-Ethylmaleimide	4×10^{-4}	30	-	-
Reserpine	10^{-5}	277	7	270
Pyrogallol	10^{-4}	142	58	84
DCNP°	10^{-4}	185	78	107
Na_2So_4	2×10^{-3}	438	33	405

Values are expressed in pmol/mg protein. Platelets (0.5 ml of EDTA-PRP) were incubated 30 min at 37°C with 10^{-6}M ^{14}C-NMN. Free and total (free plus conjugated) ^{14}C-NMN was determined by ether extraction and thin layer chromatography (Da Prada and Zürcher, 1976) of platelet perchloric acid extracts, with and without hydrolysis (20 min at 100°C; Da Prada et al., 1980). One experiment, triplicate determinations. °DCNP= 2,6-dichloro-4-nitrophenol.

garded as the major mechanism in regulation of the influx of O-methylated amines into human platelets.

In fact, platelet NMN uptake is scarcely, if at all, affected by drugs which influence active transport (oubain, cocaine and chlorimipamine) and the storage (reserpine) of the amines, but is decreased by PST inhibitors (pyrogallol and 2,6-dichloro-4--nitrophenol) and, to a greater extent, by metabolic poisons (Table 3). Further evidence that sulphoconjugation controls the diffusion of the methoxy-derivatives of CA is provided by the enhancement of NMN uptake observed in platelets incubated in the presence of sodium sulphate (Table 3). Kinetic analysis also supports this mechanism, since the K_m values for platelet uptake of NMN and MT ($\sim$ 3.0 and 0.5 uM, respectively) are similar to those of sulphoconjugation by platelet extracts (see Table 1). This is in contrast with the markedly different K_m values for platelet uptake ($\sim$ 28 uM) and sulphoconjugation (see Table 1) of other amines, such as dopamine, whose accumulation is largely dependent on active transport and granular storage (Gordon and Olverman, 1978; Da Prada and Pletscher, 1969).

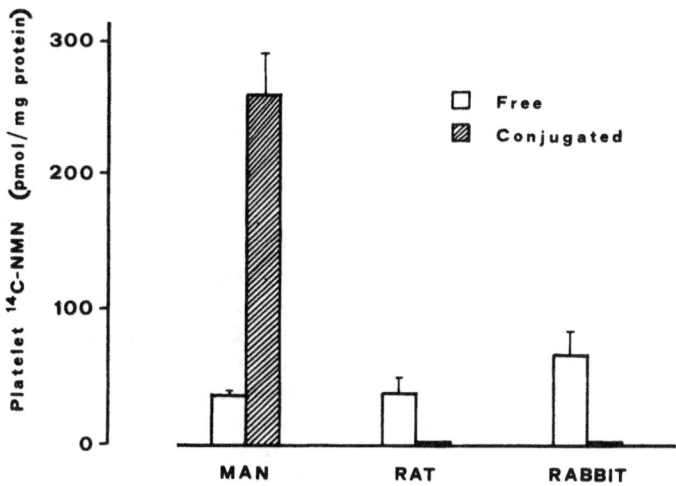

Figure 1. Free and conjugated ^{14}C-normetanephrine (NMN) in human, rat and rabbit platelets after incubation for 30 min in plasma with 10^{-6}M ^{14}C-NMN. Values are means ± SEM of 3-6 experiments.

It is worth noting that in patients with storage pool diseases the platelet, which have few and defective 5-HT organelles (for review see Da Prada et al., 1981; Lorez et al., 1979) but normal sulphoconjugation mechanisms (Da Prada, unpublished data), take up NMN in vitro to the same degree as control platelets (Picotti et al., 1981c), whereas the uptakes of 5-HT and dopamine are drastically reduced (Weiss et al., 1974; Lorez et al., 1979). Conversely, in animal platelets, which have little if any PST activity (Da Prada et al., 1980; Da Prada et al., 1981), NMN uptake is low (Figure 1) and the uptakes of 5-HT and CA are comparable to those observed in humans (Da Prada et al., 1980; Picotti, unpublished data).

The evidence that NMN uptake by platelets is closely linked to the efficiency of the sulphoconjugation mechanism raises the possibility that measurement of total NMN uptake by platelets might be a simplified way to assess variations in PST activity in intact cells. In this respect, it is worth noting that the uptake of NMN (10^{-6}M) into platelets of healthy volunteers has approximately a 2-fold intersubject variation and seems to reflect individual characteristics, since the values in the same subject were practically the same at different times over a two year period. A study comparing PST activity in platelet homogenates and NMN uptake by intact platelets from the same individuals is at present in progress in our laboratory.

An inherent difficulty in measuring endogenous monoamines in platelets is the fact that, except for 5-HT, their concentrations are very low. Only recently sensitive radioenzymatic procedures have been developed that allow the quantitative and specific determination of free CA, para-octopamine and NMN content in platelets (Da Prada and Picotti, 1979; Da Prada et al., 1980). Acid hydrolysis of human platelet extracts made it possible to establish that A, NA, dopamine, NMN and 5-HT, but not para-octopamine (which is not a preferred substrate for PST) are present in both free and conjugated forms (Da Prada et al., 1980). Among the various amines investigated, sulphated and free 5-HT have by far the highest concentrations. However, in accordance with the results of in vitro studies with platelets incubated with radioactive amines, the percent of the total endogenous platelet 5-HT in the conjugated form is lower than for CA or NMN (Table 4).

Platelets contain little, if any, biosynthetic activity for 5-HT and CA (Lovenberg et al., 1968; Marmaras and Mimikos, 1971; Solomon et al., 1970) and only a small fraction of the platelet NMN may be formed in situ by O-methylation of NA (Picotti et al., 1981b). Therefore the bulk of 5-HT, CA and NMN in the platelets are of plasma origin. It was previously shown by parallel measurements of free amines in platelets and plasma that 5-HT and CA are about 20,000 and 100-400 times more concentrated in platelets than in plasma. This concentration gradient is probably created and maintained by at least two different mechanisms, namely transfer across the cytoplasmic membrane and subcellular storage

Table 4. Content of free and conjugated amines in human platelets

	Free	Conjugated	% Conjugated
5-Hydroxytryptamine	1915 ± 377	325 ± 54	15
Dopamine	0.13 ± 0.01	0.91 ± 0.07	87
Adrenaline	0.14 ± 0.03	0.21 ± 0.06	60
Noradrenaline	1.78 ± 0.43	1.78 ± 0.64	50
Normetanephrine	1.59 ± 0.22	5.08 ± 0.27	76

Values are expressed in pmols/mg protein. Normetanephrine was determined by a modified phenylethanolamine-N-methyltransferase radioenzymatic method (Zürcher and Da Prada, in preparation). 5-Hydroxytryptamine and catecholamines were measured radioenzymatically as described by Saavedra et al.(1973) and Da Prada and Zürcher (1976). Measurements of total (free plus conjugated) amines were made in perchloric acid extracts of platelets, after hydrolysis at 100°C (Da Prada and Picotti, 1979; Da Prada et al., 1980). Means ± SEM of 3-4 subjects.

(Da Prada and Picotti, 1979; Da Prada et al., 1980).

Sulphoconjugated amines also show concentration gradients between platelets and plasma(about 150 times for NA and more than 300 times for NMN; see Da Prada et al., 1980). However, no in vitro uptake of sulphated dopamine or NMN into human platelets could be demonstrated (Picotti, unpublished data). In agreement with this finding, practically no sulphoconjugated amines have been detected in platelets of animal species, which possess very little,if any, platelet PST activity (Da Prada et al., 1980) and large amounts of sulphoconjugated amines in the plasma (Tyce et al., 1980; Kuchel et al., 1979). This supports the notion that the conjugated amines in human platelets do not originate as plasma conjugates, but are made in situ from the free amines taken up.

Therefore, the concentration gradient between platelet and plasma sulphoconjugated amines would indicate that in vivo amine sulphoconjugates formed inside the platelets do not easily cross the plasma membrane, but tend to accumulate. This agrees with in vitro uptake and release experiments which have shown that the free NMN taken up by platelets accumulates as sulphate and is only slowly liberated (the present study and Cesura et al., 1981). Interestingly enough, abnormally high levels of both free(Picotti et al., 1981a) and conjugated (data not reported) A, NA and dopamine have been observed in platelets from some patients with CA--secreting tumours.

CONCLUDING REMARKS

Since human platelets are rich in PST activity, they may function as an easily accessible tissue for studies of sulphoconjugation in man. Our findings show that in human platelets sulphoconjugation is an important mechanism for inactivating amines taken up from plasma. The extent of the conjugation appears to be dependent on the affinity of the amine for PST and on the efficiency of the uptake and storage mechanisms.

Measuring the uptake of labelled methoxy-derivatives of CA by isolated platelets should be considered as an alternative to the assay of PST activity in platelet homogenates for in vitro evaluation of the efficiency of amine sulphoconjugation by intact cells. The measurement of endogenous free and conjugated amines in platelets by radioenzymatic assay is much more laborious and time-consuming, but if done in parallel with measurement of the concentrations of these amines in plasma, it would give information about the efficiency of sulphoconjugation processes in vivo.

It has yet to be established whether or not amine sulphoconjugation in platelets reflects that in other peripheral tissues or in the brain. It should be kept in mind that animals models can not be used to obtain such information,since their platelets do not contain much, if any, PST.

ACKNOWLEDGEMENTS

    This investigation was partly supported by CNR grants nos 79.01090.83 and 80.01090.83 ("Progetto Finalizzato Medicina Preventiva"). The authors wish to thank Dr. L. Pieri for critical reading of the manuscript.

REFERENCES

    Anderson, R.J. and Weinshilboum, R.M.(1980). Phenolsulphotransferase in human tissue: radiochemical enzymatic assay and biochemical properties. Clin. Chim. Acta, 103, 79-90.

    Born, G.V.R. and Smith, J.B.(1970). Uptake, metabolism and release of ^{3}H-adrenaline by human platelets. Br. J. Pharmac., 39, 765-778.

    Cesura, A.M., Da Prada, M., Mantegazza, P. and Picotti, G.B. (1981). Uptake and sulphoconjugation of catecholamines, normetanephrine and 5-hydroxytryptamine in human platelets. Br.J.Pharmac., submitted.

    Da Prada, M. and Picotti, G.B.(1979). Content and subcellular localization of catecholamines and 5-hydroxytryptamine in human and animal blood platelets: monoamine distribution between platelets and plasma. Br. J. Pharmac., 65, 653-662.

    Da Prada, M., Picotti, G.B., Kettler, R. and Launay, J.M. (1980). Serotonin, histamine, catecholamines, normetanephrine and octopamine in blood platelets. In Platelets: Cellular Response Mechanisms and their Biological Significance,(eds. A. Rotman, F.A. Meyer, C. Gitler and A. Silberberg), Wiley, London.

    Da Prada, M. and Pletscher, A.(1969). Differential uptake of biogenic amines by isolated 5-hydroxytryptamine organelles of blood platelets. Life Sci., 8, 65-72.

    Da Prada, M., Richards, J.G. and Kettler, R.(1981). Amine storage organelles in platelets. In Platelets in Biology and Pathology, (ed. J.C. Gordon), Elsevier, Amsterdam.

    Da Prada, M. and Zürcher, G.(1976). Simultaneous radioenzymatic determination of plasma and tissue adrenaline, noradrenaline and dopamine within the fentomole range. Life Sci., 19, 1161-1174.

    De Luca, G., Barbieri, I., Ruggieri, P.and Di Giorgio, R.M. (1976). Catechol-O-methyl transferase activity in the individual human platelet population. Ital. J. Biochem., 25, 213-218.

    Eccleston, D. and Ritchie, I.M.(1973). Sulphate ester formation from catecholamine metabolites and pyrogallol in rat brain in vivo. J. Neurochem., 21,635-646.

    Foldes, A. and Meek, J.L.(1974). Occurrence and localization of brain phenolsulfotransferase. J. Neurochem., 23, 303-307.

    Gordon, J.L. and Olverman, H.J.(1978). 5-Hydroxytryptamine and dopamine transport by rat and human blood platelets. Br. J. Pharmac., 62, 219-226.

Hart, R.F., Renskers, K.J., Nelson, E.B. and Roth, J.A. 53
(1979). Localization and characterization of phenol sulphotransferase in human platelets. Life Sci., 24, 125-130.
　　Jenner, W.M. and Rose, F.A.(1973). Studies on the sulphation
of 3,4-dihydroxyphenylethylamine (dopamine) and related compounds
by rat tissues. Biochem. J., 135,109-114.
　　Kuchel, O., Buu, N.T. and Unger, T.H.(1979). Free and conjugated dopamine: physiological and clinical implications. In Advances in the Biosciences. Peripheral Dopaminergic Receptors,
(eds. J.L. Imbs and J. Schwartz), Pergamon Press, Oxford.
　　Lorez, H.P., Richards, J.G., Da Prada, M., Picotti, G.B.,
Pareti, F.I., Capitanio, A. and Mannucci, P.M.(1979). Storage
pool disease: comparative fluorescence microscopical,cytochemical
and biochemical studies on amine-storing organelles of　human
blood platelets. Brit. J. Haematol., 43, 297-305.
　　Lovenberg, W., Jéquier, E. and Sjoerdsma, A.(1968). Trypto-
phan hydroxylation in mammalian systems. Adv. Pharmacol., 6A,
21-36.
　　Marmaras, V.J. and Mimikos, N.(1971). Enzymic formation　of
serotonin in mammalian blood platelets and red cells.Experientia,
27, 196-197.
　　Mulder, G.J. and Scholtens, E.(1977). Phenol sulphotransferase and uridine diphoshate glucuronyltransferase from rat liver
in vivo and in vitro. Biochem. J., 165, 553-559.
　　Murphy, D.L.(1976). Clinical, genetic, hormonal and drug
influences on the activity of human platelet monoamine oxidase.
In Monoamine Oxidase and its Inhibition, Ciba Foundation Symposium 39, Elsevier, Amsterdam.
　　Picotti, G.B., Bondiolotti, G.P., Cesura, A.M., Ravazzani,
C., Galva, M.D. and Mantegazza, P.(1981a). Free (unconjugated)
catecholamine concentrations in platelets: biological signifi-
cance and clinical implications.In Advances in Coagulation,Plate-
let Function,Fibrinolysis and Vascular Diseases, (ed. A. Strano),
Plenum Press, New York.
　　Picotti, G.B., Cesura, A.M., Bondiolotti, G.P. and Da Prada,
M.(1981b). Uptake, storage and release of catecholamine-methoxy-
-derivatives in blood platelets. N.S. Arch. Pharmacol.,submitted.
　　Picotti, G.B., Cesura, A.M., Bondiolotti, G.P. and Galva,
M.D.(1981c).Free and conjugated normetanephrine in platelets
from patients with storage pool disease. Experientia, in press.
　　Pletscher, A.(1978). Platelets as models for monoaminergic
neurons. In Essays in Neurochemistry and Neuropharmacology,(ed.
M.B.H. Youdim), Wiley, London.
　　Pletscher, A., Burkard, W., Tranzer, J.P. and Gey, K.F.
(1967). Two sites of 5-hydroxytryptamine uptake in blood platelets. Life Sci., 6, 273-280.
　　Renskers, K.J., Feor, K.D. and Roth, J.A.(1980). Sulfation
of dopamine and other biogenic amines by human brain phenol sulfotransferase. J. Neurochem., 34, 1362-1368.

54   Saavedra, J.M., Brownstein, M. and Axelrod, J.(1973). A specific and sensitive enzymatic isotopic microassay for serotonin in tissues. J. Pharmacol. Exp. Ther., 186, 508-515.

Sandler, M. and Youdim, M.B.H.(1972). Multiple forms of mono amine oxidase: functional significance. Pharmac. Rev.,24,331-348.

Solomon, H.M., Spirt, N.M. and Abrams, W.B.(1970). The accumulation and metabolism of dopamine by the human platelet. Clin. Pharmac. Therap. 11, 838-845.

Spano, P.F. and Neff, N.H.(1971). A procedure for the simultaneous determination of dopamine, 3-methoxy-4-hydroxyphenylacetic acid, and 3,4-dihydroxyphenylacetic acid in brain.   Anal. Biochem., 42, 113-118.

Stramentinoli, G., Gualano, M., Algeri, S., De Gaetano, G. and Rossi, E.C.(1978). Catechol-O-methyl transferase (COMT) in human and rat platelets. Thrombos. Haemostas., 39, 238-239.

Tyce, G.M., Sharpless, N.S., Kerr, F.W.L. and Muenter, M.D. (1980). Dopamine conjugate in cerebrospinal fluid. J.Neurochem., 34, 210-212.

Weiss, H.J., Tschopp, T.B., Rogers, J. and Brand, H.(1974). Studies of platelet 5-hydroxytryptamine (serotonin) in storage pool disease and albinism. J. Clin. Invest., 54, 421-432.

Youdim, M.H.B. and Hefez, A.(1980). Platelet function and monoamine oxidase activity in psychiatric disorders. In Platelets: Cellular Response Mechanisms and their Biological Significance. (eds. A. Rotman, F.A. Meyer, C. Gitler and A. Silberberg), Wiley, London.

Phenol Sulfotransferase: Purification and Characterization of the Rat Brain Enzyme

Anna Baranczyk-Kuzma, Ronald T Borchardt,
Charles S Schasteen and Carol L Pinnick

Department of Biochemistry, University of Kansas, Lawrence, Kansas 66044 USA

INTRODUCTION

Phenol sulfotransferase (PST, E.C.2.8.2.1) catalyzes the transfer of a sulfate group from 3'-phospho-
adenosine 5'-phosphosulfate (PAPS) to a phenolic
acceptor substrate resulting in the formation of the
corresponding sulfate ester. This PST catalyzed reaction appears to represent a major mechanism for detoxification of endogenous and exogenous compounds bear-
ing phenolic functional groups (Roy and Trudinger,
1970; DeMeio, 1975; Jakoby, et al., 1980). Endogenous
substrates include catecholamine neurotransmitters,
e.g., dopamine (Rutledge and Hoehn, 1973), as well as
the deaminated and/or O-methylated metabolites of
catecholamines, e.g., 3-methoxy-4-hydroxyphenylglycol
(Schanberg, et al., 1968). Sulfate conjugation also
represents a major pathway for metabolism of drugs,
e.g., salicylamide, α-methyldopa (Williams, 1979).
PST activity has been isolated from various
sources including rat (Pennings, et al., 1978; Foldes
and Meek, 1973) and human brain (Renskers, et al.,
1980), and rat (Matlock and Jones, 1970; McEvoy and
Carroll, 1971; Sekura and Jakoby, 1979; Jakoby, et al.,
1980), guinea pig (Banerjee and Roy, 1966; 1968) and
rabbit liver (Hidaka, et al., 1969). Considering the
apparent differences in physiological function of the
PST's present in these tissues, they might be expected
to have quite different molecular and catalytic
characteristics. Peripheral enzymes (e.g. liver,
kidney, intestine, stomach) probably function in both
the detoxification of xenobiotic as well as endogenous

phenols, whereas brain PST probably functions primarily in the metabolism of endogenous neurotransmitters and their metabolites. In an effort to identify similarities or differences in the molecular or catalytic characteristics of PST's, we have undertaken the purification of this enzyme activity from various rat tissues including brain. In this study we report the purification of rat brain PST using an affinity chromatography system and compare its molecular and kinetic properties to the enzymes isolated from rat liver and kidney.

METHODS

PST Assays

PST activity was determined using $[^{35}S]$-3'-phospho-adenosine 5'-phosphosulfate ($[^{35}S]$-PAPS, 1.53 Ci/mmol) as the sulfate donor and a phenol, e.g. p-nitrophenol, as the sulfate acceptor. The assay mixture consisted of 10 mM phosphate buffer, pH 6.40, 0.45 μCi $[^{35}S]$-PAPS, variable amounts of the phenol, and the enzyme preparation in a total volume of 1 ml. Reaction mixtures were incubated for 5 min at 37°C. The barium hydroxide workup procedure described earlier by Foldes and Meek (1973) involves stopping the enzymatic reaction by the addition of 200 μl each of 0.1 M barium acetate, 0.1 M barium hydroxide and 0.1 M zinc sulfate. After centrifugation an aliquot of the resulting supernatant was subjected to liquid scintillation spectrometry. In the Ecteola-cellulose workup procedure 1 ml of 0.02 M ammonium bicarbonate was added to the incubation mixture prior to its application to an Ecteola cellulose column (1 x 6 cm), which had previously been equilibrated with the same buffer. The column was then eluted with a step gradient of ammonium bicarbonate (0.02 - 0.35 M) and 2 ml fractions were collected and counted by liquid scintillation spectrometry. Blanks were assayed as described above, omitting only the phenolic substrate in the incubation mixture. The enzyme specific activities are expressed in terms of nmoles of product formed per mg protein per minute. Protein concentrations were determined by the method of Lowry et al. (1951) or Bradford (1976) with crystalline bovine serum albumin as the standard protein.

Molecular Weight Determinations

The molecular weight of PST was estimated by gel filtration on a Sephadex G-100 column (1.5 x 45 cm) (Figure 4). The column was equilibrated with 5 mM

sodium phosphate buffer, pH 7.4 (1 mM DTT - 10%
glycerol) prior to addition of the protein sample. The
column was then eluted with the same buffer. The void
volume of the column was determined using Blue Dextran
2000. Various marker proteins (5 mg) [chymotrypsino-
gen A (MW 25,000), ovalbumin (MW 45,000), bovine serum
albumin (MW 69,000), aldolase (MW 158,000)] were used
to calibrate the column.

Polyacrylamide Gel Electrophoresis
      Polyacrylamide gel electrophoresis of the purified
PST samples was carried out using the procedures
described by Davis (1964). Electrophoresis was done
with 7.5% crosslinked gels in tris-glycine buffer, pH
8.3 at 4°C. A current of 2 mA per gel was applied for
15 min, followed by 6 mA per gel for about 60 min.
The gels were fixed and stained with 0.005% Coomassie
Blue G-250 in 10% isopropanol and 10% acetic acid. In
unstained gels PST activity was determined in 2 mm gel
slices after first eluting the enzyme with 10 mM
sodium phosphate buffer, pH 6.4.

PST Purification
      Step 1. Subcellular fractionation. ARS Sprague
Dawley male rats (150-175 g) were killed by decapitation; the tissues removed, washed with cold 5 mM sodium
phosphate buffer, pH 6.4 (1 mM DTT - 10% glycerol),
and homogenized in 3 volumes of the same buffer using
a glass homogenizer equipped with a loose-fitting
motor-driven Teflon pestle. All subsequent steps are
carried out at 4°C. The homogenates were centrifuged
for 20 min at 9,000 g. The supernatant liquid was
decanted; and the resulting pellet was again homogenized
in the same buffer, centrifuged for 20 min at 9,000 g,
and the resulting supernatant combined with the first
extract. The combined supernatants were centrifuged
at 100,000 g for 90 min.
      Step 2. Ammonium sulfate fractionation. The
cytosolic fractions were further subfractionated using
ammonium sulfate. The precipitates at 35-55% salt for
liver, 0-55% salt for kidney, and 40-60% salt for brain
were collected by centrifugation at 14,000 g for 20
min. The precipitates were dissolved in 5 mM phosphate
buffer, pH 7.4 (1 mM DTT - 10% glycerol) and dialyzed
for 20 hr against the same buffer.
      Step 3. DEAE-cellulose chromatography. The
dialyzed protein solutions were applied to DEAE-
cellulose columns (2 x 30 cm) equilibrated with 5 mM
sodium phosphate buffer, pH 7.4 (1 mM DTT - 10% glycerol). The columns were then eluted with a linear

KCl gradient (0-500 mM) (Figure 1), collecting 5 ml
fractions. The fractions containing PST activity were
pooled and concentrated by ultrafiltration through an
Amicon PM-10 membrane. The samples were dialyzed
against 5 mM sodium phosphate, pH 7.4 (1 mM DTT - 10%
glycerol) in preparation for affinity chromatography.

Step 4. Affinity chromatography. PST's purified
through Step 3 were applied to p-hydroxyphenylacetic
acid-agarose affinity columns (2.8 x 28 cm) (Borchardt
and Schasteen, unpublished results), which were
equilibrated with 5 mM sodium phosphate, pH 7.4 (1 mM
DTT - 10% glycerol). In general, a major protein peak
lacking PST activity was eluted with the 5 mM phosphate
buffer, after which a linear phosphate buffer, pH 7.4
gradient (5-200 mM) was used to elute the PST (Figure
3). The fractions containing PST activity were pooled,
concentrated by ultrafiltration and dialyzed for 20 hr
against 5 mM sodium phosphate buffer, pH 7.4 (1 mM DTT
- 10% glycerol).

Step 5. Sephadex G-100 chromatography. PST's
purified through Step 4 were applied to Sephadex G-100
columns (1.5 x 45 cm), which were previously equilibrated with the 5 mM sodium phosphate buffer, pH 7.4
(1 mM DTT - 10% glycerol). The same buffer was used
to elute the PST activity (Figure 4). The fractions
containing PST activity were pooled, concentrated and
stored at -80°C.

RESULTS

Purification of Brain PST
    The general scheme developed for the purification
of rat brain PST is shown in Table 1. The initial
steps in the purification scheme are modeled after the
procedures described earlier by Foldes and Meek (1973),
that being differential centrifugation followed by
ammonium sulfate fractionation. The subcellular
distribution of PST was found to be primarily cytosolic with little or no detectable activity in the
microsomal or mitochondrial fractions. In the salt
fractionation procedure a maximum increase in specific
activity was obtained using a 40-60% ammonium sulfate
fractionation step. After dialysis the salt fractionated enzyme was applied to a DEAE-cellulose column.
As shown in Figure 1 the PST activity was eluted using
a linear KCl gradient (0-500 mM). A symmetrical peak
of PST activity was observed without any indication
of isozymic forms similar to those reported earlier
for the rat liver enzyme (Sekura and Jakoby, 1970;
Jakoby et al., 1980).

TABLE 1. Purification of rat brain PST

Step	Fraction [a]	Total Activity [b] (nmol/min)	Specific Activity [b] (nmol/min/mg protein)	Purification
	9,000 g Supernatant	5.5	3.3×10^{-3}	—
1	100,000 g Supernatant	5.3	4.8×10^{-3}	1.5
2	(NH$_4$)$_2$SO$_4$ Fractionation	3.4	1.4×10^{-2}	4.2
3	DEAE-Cellulose Chromatography	3.2	7.2×10^{-2}	22
4	Affinity Chromatography	2.2	5.5×10^{-1}	167
5	Sephadex G-100 Chromatography	1.8	2.7	450

[a] See text for details concerning steps of purification.

[b] PST activity was assayed using p-nitrophenol and [35S]-PAPS as substrates and the barium hydroxide workup procedure described by Foldes and Meek (1973).

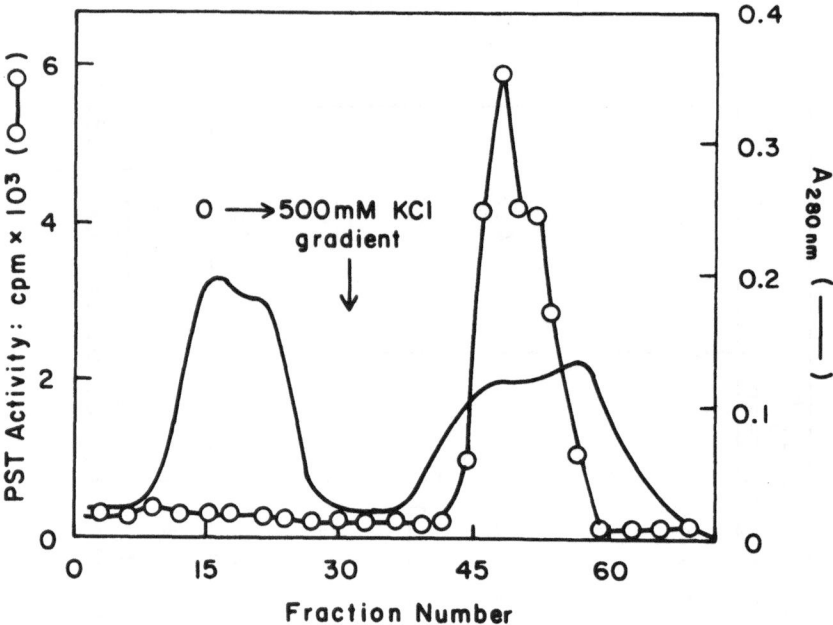

Figure 1. DEAE-cellulose chromatography of brain PST. Brain PST (242
mg) purified through the ammonium
sulfate fractionation step (40-60%
salt) was applied to a DEAE-cellulose
column (2 x 30 cm) in 5 mM sodium
phosphate buffer, pH 7.4 (1 mM DTT -
10% glycerol). PST activity was
eluted with a linear KCl gradient
(0-500 mM) in the same buffer, collecting 5 ml fractions. The fractions were assayed for PST activity
using p-nitrophenol (5 µM) as substrate and the barium hydroxide workup procedure described by Foldes and
Meek (1973).

    The brain PST was further purified using an affin-
ity chromatography system consisting of a p-hydroxyphenylacetic acid-agarose conjugate as shown in Figure
2. In earlier studies Borchardt and Schasteen (unpublished data) have shown that the most effective
affinity system for purification of the rat liver
enzyme consists of the ligand separated from the
insoluble matrix by a spacer of 25 A°. As shown in

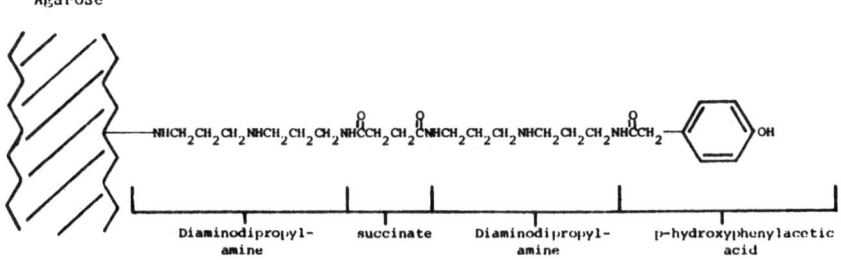

Figure 2. Structure of the p-hydroxyphenylacetic acid-agarose conjugate used for affinity chromatography of PST.

Figure 2 the spacer arm consists of repeating units of diaminopropylamine and succinic acid. The ability of the p-hydroxyphenylacetic acid-agarose conjugate to bind brain PST is shown in Figure 3. The DEAE-cellulose purified enzyme was applied to the affinity column in 5 mM phosphate buffer, pH 7.4. Under these conditions all of the PST activity was retained by the column. No detectable PST activity could be eluted even after extensive washing with the same buffer. The PST activity could be eluted from the column with a linear phosphate gradient (5-200 mM). Specific interaction of PST and the matrix-bound ligand was apparent since unsubstituted agarose did not retain the PST activity in a similar manner. This step in purification affords approximately an 8-fold increase in the specific activity of the enzyme.

Brain PST was further purified by chromatography on Sephadex G-100 as shown in Figure 4a. The pooled brain enzyme from the Sephadex G-100 chromatography

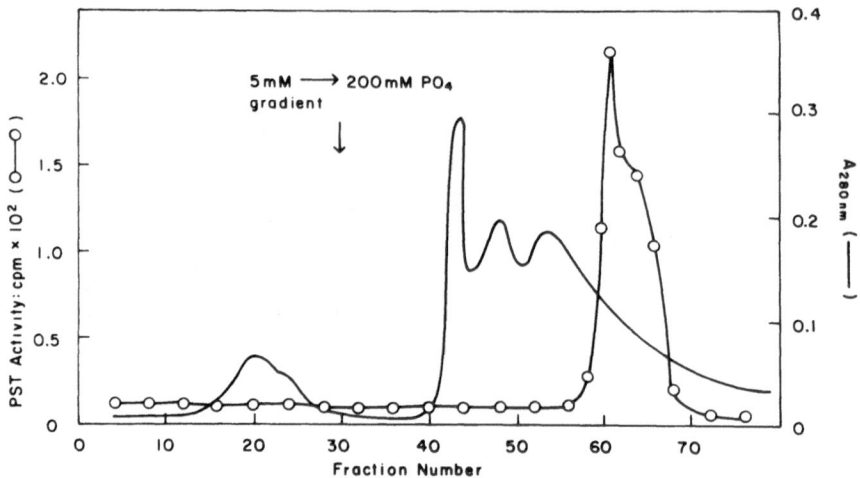

Figure 3. Affinity chromatography
of brain PST. Brain PST (44 mg)
purified through the DEAE-cellulose
step was applied to a p-hydroxyphenylacetic acid-agarose affinity
column (2.8 x 28 cm). The column
was initially eluted with 5 mM
sodium phosphate buffer, pH 7.4
(1 mM DTT - 10% glycerol), followed
by a linear phosphate gradient
(5-200 mM), collecting 5 ml fractions. PST activity was determined
as described in Figure 1.

step had a specific activity of 2.7 nmol/mg protein/
min, which represented a 450-fold purification over the
activity in the cell homogenate.

Properties of Brain PST
    By calibrating the Sephadex G-100 column with
proteins of known molecular weights we were able to
estimate a molecular weight of 68,000 for brain PST.
This molecular weight was identical to that observed
for the liver and kidney enzymes isolated by the same
procedures (Kuzma, et al., unpublished data). When

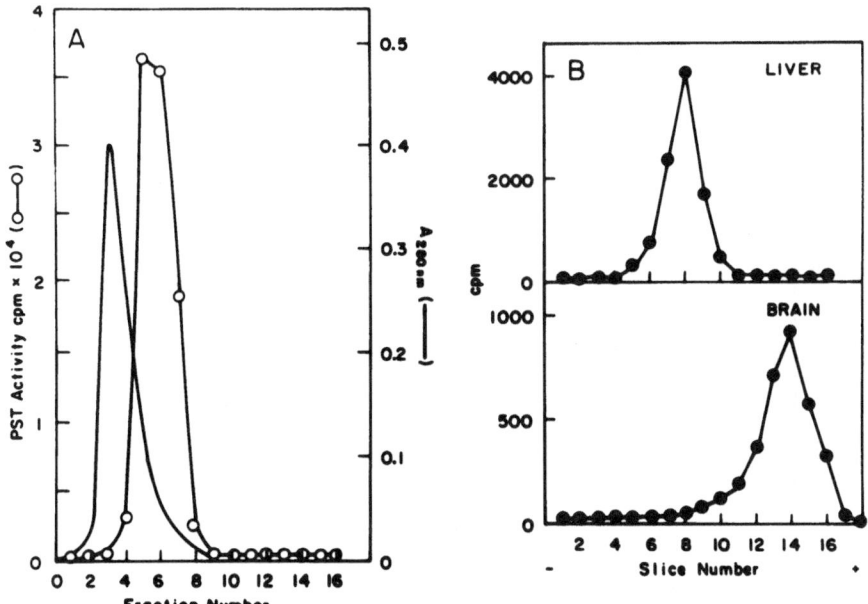

Figure 4a. Sephadex G-100 gel
filtration of brain PST. Brain PST
(4 mg) purified through the affinity
chromatography step was applied to
a Sephadex G-100 column (1.5 x 45 cm).
The column was eluted with 5 mM
sodium phosphate buffer, pH 7.4
(1 mM DTT - 10% glycerol),
collecting 2 ml fractions. PST
activity was determined as described
in Figure 1.

Figure 4b. Polyacrylamide gel
electrophoresis pattern for rat
brain and liver PST. Enzymes
were purified through the affinity
chromatography step. Electrophoresis
was carried out using 7.5% gels in
pH 8.3 tris-glycine buffer. The
tubes were subjected to 2 mA for
15 min followed by 6 mA for 60 min.
PST activity was determined in 2 mm
gel slices, first eluting the enzyme
with 10 mM sodium phosphate buffer,
pH 6.4.

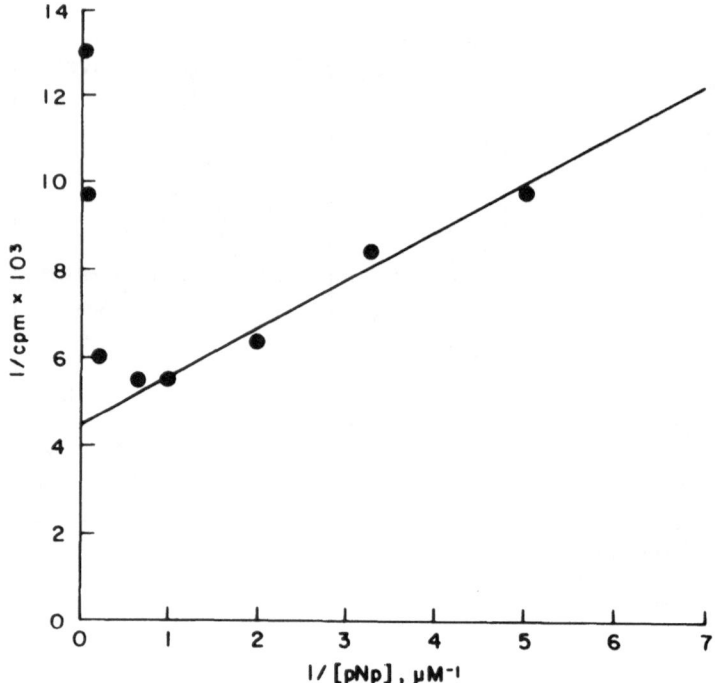

Figure 5. Lineweaver-Burke plot
for brain PST activity with p-
nitrophenol as the variable substrate. The barium hydroxide workup
procedure of Foldes and Meek (1973)
was used to isolate the [35S]-
sulfated product.

this purified brain PST was subjected to polyacrylamide
gel electrophoresis, then the gels sliced and assayed
for PST activity, a single symmetrical peak of enzyme
activity was observed (Figure 4b). It is interesting
to note that the brain PST exhibits a quite different
mobility on gel electrophoresis than the rat liver PST.
     The pH profile for this enzyme activity was deter-
mined using p-nitrophenol and [35S]-PAPS as substrates.
In the pH range of 4 to 9 the maximum activity was
observed at pH 5.6.
     The substrate kinetic properties for the purified
brain enzyme were also determined. The Km for PAPS
was found to be 0.29 ± 0.05 μM and that for p-nitrophenol was estimated to be 0.26 ± 0.02 μM. The low Km
value for p-nitrophenol is particularly interesting
since the Km values observed for the peripheral enzymes
(liver, kidney and stomach) are approximately 9-16 μM

(Kuzma, et al., unpublished data). Consistent with
the results reported earlier by Foldes and Meek (1973)
we observed that the brain enzyme was inhibited by
increasing concentrations of the substrate p-nitrophenol (Figure 5). This inhibitory effect of high
concentrations of the phenolic substrate appears to
be a characteristic unique to the brain enzyme since
it was not observed with the liver, kidney or stomach
enzymes (Kuzma, et al., unpublished data).

Mulder and Scholtens (1977) have reported that 2,6-
dichloro-4-nitrophenol is a potent and selective inhibitor of sulfation in vivo and in vitro. Therefore,
the inhibitory effect of 2,6-dichloro-4-nitrophenol
on brain PST was determined. This compound was found
to be a noncompetitive inhibitor of brain PST when
p-nitrophenol was the variable substrate. The observed
inhibition constants were K_{ii} = 0.11 ± 0.03 µM and
K_{is} = 0.8 ± 0.01 µM. In contrast, the compound is a
competitive inhibitor of the liver, kidney and stomach
enzymes with Ki's of approximately 0.25 to 0.60 µM
(Kuzma, et al., unpublished results). In earlier
studies Borchardt and Schasteen (1977) found that rat
liver PST has an essential arginyl residue at its
active site. This arginine residue appears to be
involved in binding PAPS. Similarly, in this study
we have observed that brain PST can be inactivated by
phenylglyoxal (data not shown), an arginine modifying
reagent; and that this inactivation can be prevented
by inclusion of PAPS.

Since we had encountered difficulty with the
barium hydroxide workup procedure for the PST assay
described earlier by Foldes and Meek (1973), we have
developed an alternative workup procedure which allows
for the accurate estimation of the amount of sulfated
esters being produced from a variety of structurally
related phenols. This new workup procedure involves
chromatography of the PST incubation mixture on an
Ecteola-cellulose column eluting with increasing concentrations of ammonium bicarbonate (0.02-0.35 M).
Ecteola-cellulose is an anion exchange resin which
was used earlier by Balasubramanian, et al. (1967)
and Shoyab, et al. (1972) for the purification of PAPS.
In Figure 6 is shown the elution pattern of a PST
reaction mixture containing [35S]-PAPS and [35S]-p-
nitrophenol sulfate. The base line separation of the
labeled substrate ([35S]-PAPS) from the labeled product ([35S]-p-nitrophenol sulfate) allows for an
accurate estimation of the enzymatic velocity. A
chromatogram of an incubation mixture lacking the
phenolic substrate, e.g. p-nitrophenol, served as the

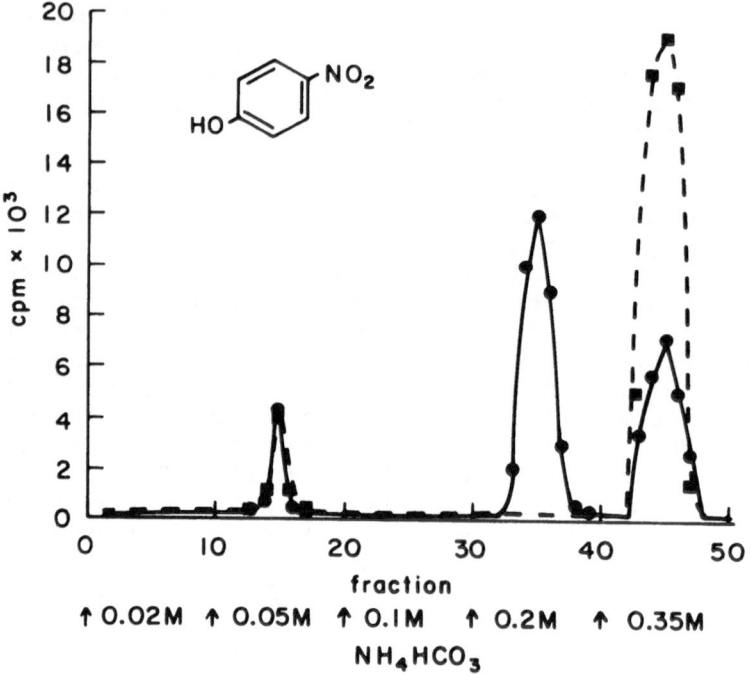

Figure 6. Ecteola-cellulose chromatography of the brain PST reaction
products. Brain PST was incubated
for 5 min at 37°C with 5 μM p-nitrophenol, [^{35}S]-PAPS (1.53 Ci/mmol)
and 10 mM sodium phosphate buffer,
pH 6.4. An aliquot of the reaction
mixture was applied to an Ecteola
cellulose column (1 x 6 cm) (•-•).
The column was washed with increasing concentrations of ammonium
bicarbonate (0.02-0.35 M),
collecting 2 ml fractions. Radioactivity was determined as described
in the text. □-□ was the same
reaction mixture minus p-nitrophenol.

blank (Figure 6). The mobilities of the sulfated
phenols on Ecteola-cellulose depend upon their anion
character, but without exception we were able to
separate such sulfate esters from PAPS using this anion
exchange resin.
     As shown in Table 2 the amount of [^{35}S]-p-nitro-
phenol sulfate generated in a PST incubation mixture
can be accurately estimated using either the barium

TABLE 2. Substrate specificity of brain PST determined using the barium hydroxide workup versus the Ecteola-cellulose workup. [a]

Substrates	% Activity	
	Barium hydroxide assay [b]	Ecteola-cellulose assay [c]
p-nitrophenol	100	100
3,4-dihydroxyphenylacetic acid	0	325
3,4-dihydroxymandelic acid	31	193
homovanillic acid	44	274
3,4-dihydroxyphenylglycol	84	201
3-methoxy-4-hydroxyphenylglycol	110	250
homovanillic alcohol	137	256
salicylamide	40	33
vanillin	117	153
vanillyl alcohol	23	62
p-hydroxybenzaldehyde	73	116

[a] PST activity was determined using the indicated phenolic substrates (5 μM), [35S]-PAPS (1.53 Ci/mmol) and 10 mM sodium phosphate buffer, pH 6.4. Incubations were carried out at 37° for 5 min.

[b] The barium hydroxide workup consisted of precipitating the unreacted [35S]-PAPS with barium hydroxide and zinc sulfate, leaving the [35S]-labelled sulfated product in solution. An aliquot of the resulting supernatant is then counted.

[c] In the Ecteola-cellulose workup the reaction mixture is chromatographed on Ecteola cellulose eluting with an ammonium bicarbonate gradient (0.02-0.35 M) as described in the text.

hydroxide workup procedure or the Ecteola-cellulose
workup procedure. The barium salt of [35S]-p-nitro-
phenol sulfate is very water soluble; subsequently
little is lost in this workup procedure. In contrast,
highly anionic [35S]-phenol sulfates are precipitated
in the standard barium hydroxide workup procedure, and
subsequently the corresponding phenols, e.g., 3,4-
dihydroxyphenylacetic acid, 3,4-dihydroxymandelic acid,
homovanillic acid, appear to be poor substrates for
PST. As shown in Table 2, when the Ecteola-cellulose
workup procedure is used these phenols appear to be
excellent substrates for brain PST. Even neutral
phenols, such as 3,4-dihydroxyphenylglycol, 3-methoxy-
4-hydroxyphenylglycol and homovanillic alcohol, show
better substrate properties for PST when the Ecteola
cellulose workup procedure was used instead of the
barium hydroxide procedure (Table 2). Therefore, the
Ecteola-cellulose workup procedure allows for a more
accurate estimation of a phenol's substrate properties
for PST.

Using this Ecteola-cellulose workup procedure, we
have determined the substrate specificity of the rat
brain, kidney and liver enzymes (Table 3). Of particular interest was the observation that the deaminated
and/or O-methylated metabolites of catecholamines
are extremely good substrates for brain PST. The same
phenols show little or no substrate activity with the
liver and kidney enzymes. This high specificity for
the deaminated and/or O-methylated metabolites of
catecholamines appears to be consistent with the physiological function of the brain enzyme. Under the
conditions of the assay (pH = 6.4), the catecholamines,
e.g., epinephrine, norepinephrine, dopamine, or their
O-methylated metabolites, e.g., normetanephrine, metanephrine, did not exhibit good substrate properties.
Other phenolic amines, e.g., serotonin, tyramine, were
also poor substrates for this enzyme activity. The
poor substrate properties of these amines probably
result because of the pH 6.4 assay conditions, since
Foldes and Meek (1973) have reported earlier that the
pH optimums for these amines with rat brain PST was
about pH 9.0.

As shown in Table 3, several xenobiotics were also
evaluated as substrates for rat liver, kidney and brain
PST's. Salicylamide and vanillin in particular appear
to be good substrates for all of the enzymes tested.

DISCUSSION

Rat brain PST was partially purified earlier by

TABLE 3. Substrate specificity of rat liver, kidney and brain PST.[a]

Substrates	% Activity		
	Liver	Kidney	Brain
p-nitrophenol	100% (11 nmol/mg/min)	100% (1.6 nmol/mg/min)	100% (0.072 nmol/mg/min)
3,4-dihydroxyphenylacetic acid	0	0	325
3,4-dihydroxymandelic acid	0	0	193
homovanillic acid	0	0	274
3,4-dihydroxyphenylglycol	0	0	201
3-methoxy-4-hydroxyphenylglycol	35	0	250
homovanillic alcohol	34	0	256
DL-epinephrine	24	0	30
DL-norepinephrine	7	0	0
tyramine	27	16	22
DL-metanephrine	0	0	0
DL-normetanephrine	0	0	0
dopamine	13	0	0
serotonin	0	0	0
salicylamide	102	54	33
vanillin	77	56	153
vanillyl alcohol	0	0	62
p-hydroxybenzaldehyde	0	0	116

[a] PST activity was determined using the indicated phenol (liver, kidney, 5 μM) and [35S]-PAPS. The reaction mixture was chromatographed on an Ecteola cellulose column eluting with an ammonium bicarbonate gradient (0.02-0.35 M) as described in the text.

Foldes and Meek (1973) and Pennings et al., (1978).
The Foldes and Meek (1973) purification scheme involved
differential centrifugation, ammonium sulfate fractionation, alumina adsorption and DEAE-cellulose chromatography, producing a 30-fold increase in specific activity. In the current studies we have combined classical
protein purification techniques, e.g., ion exchange and
gel filtration chromatography, with the technique of
affinity chromatography and achieved a 450-fold purification of the rat brain enzyme. The most effective
affinity system developed in our laboratory (Borchardt
and Schasteen, unpublished data) for purification of
PST was a p-hydroxyphenylacetic acid-agarose conjugate
with the ligand separated from the insoluble matrix by
a spacer of approximately 25 A°. Increases or
decreases in the length of the spacer arm resulted in
decreases in the effectiveness of the system for
purification of PST (Borchardt and Schasteen, unpublished results).

Several commercially available affinity systems
have recently been used for purification of PST's.
Affi-Gel Blue and ATP-agarose were used to purify
several aryl sulfotransferases from rat liver (Sekura
and Jakoby, 1979; Jakoby, et al., 1980). Both in the
current study of brain PST and the study reported
earlier by Sekura and Jakoby (1979) on rat liver PST,
the use of affinity chromatography has afforded enzymes
of much higher purity than procedures published
earlier (Foldes and Meek, 1973; Pennings, et al., 1978;
Matlock and Jones, 1970; McEvoy and Carroll, 1971).
These more highly purified preparations of PST should
permit elucidation of the molecular and catalytic
characteristics of aryl sulfotransferases.

The properties of rat brain PST reported in this
study are in general consistent with those reported
earlier by Foldes and Meek (1973) who used a less
highly purified enzyme preparation. Some very interesting differences, however, were observed between our
purified brain PST and the enzyme isolated by the same
techniques from rat liver (Borchardt, et al., unpublished results). Of particular interest were the differences in mobility on gel electrophoresis suggesting
different molecular structures, and the sensitivity of
the brain enzyme to substrate inhibition suggesting
differences in the catalytic mechanisms. Several
similarities between the rat liver and brain enzymes
were observed including the molecular weights (approximately 68,000), and their sensitivities to the inhibitors 2,6-dichloro-4-nitrophenol and phenylglyoxal.
The inhibition by phenylglyoxal suggests the enzymes

have arginyl residues at their active sites (Borchardt
and Schasteen, 1977).

In order to accurately determine the substrate
specificity of the brain PST, we have developed an
alternative workup procedure for the PST assay described earlier by Foldes and Meek (1973). Earlier
workers (Foldes and Meek, 1973; Chen, et al., 1978)
suggested that the standard barium hydroxide workup
procedure for the PST assay is not effective with sulfate esters of certain acidic compounds. Our own
experience verified these observations and further
suggested that the solubility of the barium salts of
phenol sulfates vary greatly depending upon the
structural features of the molecules. Unless these
solubility properties were corrected for the resulting
data from PST assays could give misleading results concerning the substrate properties of phenols. The basis
of the barium hydroxide workup procedure is the "relative" insolubility of the barium salts of PAPS and
inorganic sulfate and the "relative" solubility of the
barium salts of phenol sulfates in water. The major
problem with this workup procedure is the lack of
information concerning the exact solubility properties
of the barium salts of phenol sulfates and the neces-
sity, therefore, to generate such data in order to
obtain an accurate assessment of the substrate properties of a phenol for PST. In the Ecteola-cellulose
workup procedure described in this paper, it is not
necessary to know those physical-chemical parameters
in order to accurately assess the ability of a phenol
to serve as a substrate for PST.

Utilizing the Ecteola-cellulose assay we observed
some rather significant differences in the substrate
specificity of the rat liver and brain enzymes. The
brain enzyme, but not the liver or kidney enzyme,
has a high affinity for deaminated and/or O-methylated
metabolites of catecholamines. Both neutral and acidic
catecholamine metabolites were found to be good substrates for brain PST. In contrast, the amines themselves and their O-methylated metabolites were found
to be poor substrates for the brain enzyme as well as
liver and kidney PST. All three enzyme preparations
were able to readily sulfate xenobiotic phenols, such
as salicylamide or vanillin. The specificity of the
brain PST for the neutral and acidic catecholamine
metabolites correlates well with its probable physiological function, that being to catalyze the formation
of highly ionized organic sulfates of catecholamine
metabolites that are more water soluble than the
parent compounds and thereby facilitate excretion.

Balasubramanian, A. S., Spolter, L., Rice, L.I.,
    Sharon, J. B. and Marx, W. (1967). Preparation of
    3'-Phosphoadenosyl Sulfate in Substrate Quantities
    using Mastocytoma Enzymes. Anal. Biochem., 21,
    22-23.
Banerjee, R. K. and Roy, A. B. (1966). The Sulfotrans-
    ferases of Guinea Pig Liver. Mol. Pharmacol., 2,
    56-66.
Banerjee, R. K. and Roy, A. B. (1968). Kinetic Studies
    of Phenol Sulfotransferase Reaction. Biochim.
    Biophys. Acta, 151, 573-587.
Borchardt, R. T. and Schasteen, C. S. (1977). Phenol-
    Sulfotransferase Inactivation by 2,3-Butanedione
    and Phenylglyoxal: Evidence for an Active Site
    Arginyl Residue. Biochem. Biophys. Res. Commun.,
    78, 1067-1073.
Bradford, M. M. (1976). A Rapid and Sensitive Method
    for the Quantitation of Microgram Quantities of
    Protein Utilizing the Principle of Protein-Dye
    Binding. Anal. Biochem., 72, 248-254.
Chen, L. J., Imperto, T. J. and Bolt, R. J. (1978).
    Enzymatic Sulfation of Bile Salt Sulfotransferase
    from Rat Kidney. Biochim. Biophys. Acta, 522,
    443-451.
Davis, B. J. (1964). Disc Electrophoresis - II.
    Method and Application to Human Serum Proteins.
    Ann. N. Y. Acad. Sci., 121, 404-427.
DeMeio, R. H. (1975). Sulfate Activation and Transfer.
    In Metabolic Pathways: Third Edition, (ed., D. M.
    Greenberg), Academic Press, New York, Vol. 7, p.
    287-358.
Foldes, A. and Meek, J. L. (1973). Rat Brain Phenol-
    sulfotransferase - Partial Purification and Some
    Properties. Biochim. Biophys. Acta, 327, 365-373.
Hidaka, H., Nagatsu, T. and Yagi, K. (1969). Forma-
    tion of Serotonin O-Sulphate by Sulphotransferase
    of Rabbit Liver. Biochim. Biophys. Acta, 177,
    354-357.
Jakoby, W. B., Sekura, R. D., Lyon, E. S., Marcus,
    C. J. and Wang, J. L. (1980). Sulfotransferases.
    In Enzymatic Basis of Detoxication, (ed., W. B.
    Jakoby), Academic Press, New York, p. 199-228.
Lowry, O. H., Rosenbrough, N. J., Farr, A. L. and
    Randall, R. J. (1951). Protein Measurement with
    the Folin Phenol Reagent. J. Biol. Chem., 193,
    265-273.
Matlock, P. and Jones, J. F. (1970). Partial Purifica-
    tion and Properties of an Enzyme from Rat Liver

that Catalyzes the Sulphation of L-Tyrosine Derivatives. Biochem. J., _116_, 797-803.

McEvoy, F. A. and Carroll, J. (1971). Purification from Rat Liver of an Enzyme that Catalyzes the Sulphurylation of Phenols. Biochem. J., _123_, 901-906.

Mulder, G. and Scholtens, E. (1977). Phenol Sulphotransferase and Uridine Diphosphate Glucuronyltransferase from Rat Liver _in vivo_ and _in vitro_. Biochem. J., _165_, 553-559.

Pennings, E. J. M., Vrielink, R. and Van Kempen, G. M. J. (1978). Kinetic and Mechanism of the Rat Brain Phenol Sulfotransferase Reaction. Biochem. J., _173_, 299-307.

Renskers, K. J., Feor, K. D. and Roth, J. A. (1980). Sulfation of Dopamine and Other Biogenic Amines by Human Brain Phenol Sulfotransferase. J. Neurochem., _34_, 1362-1368.

Roy, A. B. and Trudinger, P. A. (1970). The Biochemistry of Inorganic Compounds of Sulfur. Cambridge University Press, London.

Rutledge, C. O. and Hoehn, M. M. (1973). Sulfate Conjugation and L-DOPA Treatment of Parkinsonian Patients. Nature, _244_, 447-450.

Schanberg, S. M., Breese, G. R., Schildkraut, J. J., Gordon, E. K. and Kopin, I. J. (1968). 3-Methoxy-4-hydroxyphenylglycol sulfate in Brain and Cerebrospinal Fluid. Biochem. Pharmacol., _17_, 2006-2008.

Sekura, R. D. and Jakoby, W. B. (1979). Phenol Sulfotransferases. J. Biol. Chem., _254_, 5658-5663.

Shoyab, M., Su, L. Y. and Marx, W. (1972). Purification and Properties of ATP-Sulfurylase from Furth Mouse Mastocytoma. Biochim. Biophys. Acta, _258_, 113-124.

Wengle, B. (1964). Studies on Ester Sulphates 16. Use of [35S]-Labelled Inorganic Sulphate for Quantitative Studies of Sulphate Conjugation in Liver Extracts. Acta Chem. Scand., _18_, 65-76.

Williams, R. T. (1979). The Metabolism of Phenols - The Sulphate Conjugation. In Detoxication Mechanisms, Wiley, New York, p. 279-284.

ACKNOWLEDGEMENT

The authors gratefully acknowledge support of this project by grants from the National Institutes of Health (GM-22357) and the American Heart Association, Kansas Affiliate.

Characterization of Human Brain Phenol Sulfotransferase

Jerome A Roth, Jennifer Rivett and Kevin Renskers

Department of Pharmacology and Therapeutics, SUNY at Buffalo,
Buffalo, New York, 14214 USA

The biogenic amine neurotransmitters are inactivated in the central nervous system by several enzymatic reactions including deamination by monoamine oxidase (MAO), 0-methylation by catechol-0-methyltransferase (COMT) and 0-sulfation by phenol sulfotransferase (PST). The properties and location of both MAO and COMT have been extensively studied in human brain whereas the enzyme or enzymes involved in sulfate conjugation have received relatively little attention. Recent evidence (Anderson and Weinshilboum, 1980; Tyce et al., 1980; Renskers et al., 1980) has suggested that sulfate conjugation may contribute significantly to the overall inactivation of the putative amine neurotransmitters, especially dopamine. However, quantitative information as to the magnitude of this enzymatic process in human brain has not been assessed.

Several papers have previously reported that catecholamines undergo extensive sulfate conjugation in man. As early as 1940, Richter demonstrated that almost 65% of orally administered epinephrine was recovered in the urine as the sulfate ester. Since then numerous papers have appeared in the literature which have demonstrated that the catecholamines are extensively conjugated in the periphery of humans (Haggendal, 1963; Kahane et al., 1967; Goodall and Alton, 1972; Rutledge and Hoehn, 1973; Jenner and Rose, 1974; Bronough et al., 1975; Hart et al., 1979). A recent paper by Johnson et al. (1980) has revealed that the endogenous catecholamines in the human circulation exist primarily as the sulfate ester. They reported that approximately 73% of the total norepinephrine and 99% of the total dopamine in plasma of normotensive subjects is conjugated. These studies clearly indicate that sulfation of both epinephrine and norepinephrine represents a major catabolic pathway for the catecholamine neuro-

transmitters, at least, in the periphery. Prior studies by
Richter and MacIntosh (1941) have demonstrated that esterifica-
tion of epinephrine reduces its pressor activity and accordingly,
these investigators have proposed that sulfation is a major
mechanism for inactivation of this amine.

Recent studies by Tyce and coworkers (Tyce et al., 1980)
which demonstrate the presence of dopamine sulfate in the cerebrospinal fluid of Parkinson patients suggests that this esteri-
fication reaction may also occur in brain. In support of this
premise, studies in our laboratory (Renskers et al., 1980)
reveal that both dopamine and norepinephrine undergo sulfate
conjugation in subcellular preparations of human brain. The
enzyme involved in this reaction, phenol sulfotransferase,
utilizes 3'-phosphoadenosine-5'-phosphosulfate (PAPS) as the
sulfate donor and is found primarily in the soluble fraction of
brain homogenates. As a continuation of our previous work, we
will present in this chapter some important new findings as to
the contribution of PST, relative to that of MAO and COMT, to
the inactivation of catecholamines in human brain.

METHODS

PST was assayed by the method of Foldes and Meek (1973)
utilizing 35S-labeled PAPS. When necessary, 1 mM of the MAO
inhibitor, pargyline, was added to prevent deamination of dopamine. MAO A and B activity was measured as described previously
using dopamine as substrate (Roth and Feor, 1978). Type B MAO
activity was based on the dopamine deaminating activity in each
brain preparation that was insensitive to the selective type A
MAO inhibitor, clorgyline. The difference between the total MAO
activity and that which was insensitive to clorgyline represented type A MAO activity. Microsomal and soluble COMT activity
was measured with ^{3}H-dopamine as described in a recent publication from this laboratory (Roth, 1980). Unless otherwise
specified, all kinetic constants for MAO type A and B activity
were obtained using mitochondrial preparations of human brain.
Kinetic constants for microsomal and soluble COMT activity were
determined using isolated preparations of the appropriate
subcellular fractions.

RESULTS AND DISCUSSION

Biochemical Characterization of PST
In a previous communication from this laboratory (Renskers
et al., 1980), it was demonstrated that human brain PST has the
capacity to esterify a variety of biogenic amines and their
deaminated and O-methylated metabolites. The Km values of these

compounds are presented in Table 1. As the data clearly demonstrate, human brain PST has by far the highest affinity for the catecholamines and their O-methylated metabolites. The absence of a meta substituent on the phenolic ring, whether it be a hydroxyl or a methoxy group, has the effect of greatly decreasing the binding affinity. This is seen by the fact that PST has reduced affinities for the phenolic amines, tyramine and octopamine, as compared to their meta-substituted analogs. Both tyramine and octopamine as well as their respective meta-methoxy derivatives can only be sulfoconjugated in the hydroxylated para position of the aromatic ring whereas dopamine can be esterified in either position. Prior studies have revealed that conjugation of the meta position predominates with human brain PST. Thus, it is surprising that the meta-unsubstituted phenolic amines have a lower affinity for PST when the size and properties of substitutents (hydroxyl versus methoxy) on this ring position apparently have less of an effect on the binding affinities of these compounds.

TABLE 1

Km values for binding of biogenic amines and metabolites to human brain phenolsulfotransferase*

Substrate	Km (μM)
3-Methoxytyramine	1.4
Dopamine	2.8
Normetanephrine	8.7
Epinephrine	11
Norepinephrine	22
Tyramine	79
3-Methoxy-4-hydroxyphenylethyleneglycol	320
Octopamine	460
3,4-Dihydroxyphenylacetic acid	940

*PST activity was measured in 10 mM potassium phosphate buffer, pH. 7.2, containing 1 μM PAPS and 1 mM pargyline using a 100,000 g (60 mins) supernatant

Although the catechol structure facilitates binding of these compounds to the enzyme, this structural feature in itself is clearly not the only chemical determinant which regulates binding to PST. Comparing the Km values of dopamine and its O-methylated derivative to that of norepinephrine and normetanephrine it can be seen that the presence of the B-hydroxyl group on the aliphatic amine side chain greatly reduces the binding affinities of these catecholamines to PST. The data further

indicate that the amine group also influences binding since the
deaminated products of dopamine and normetanephrine have reduced
affinities for this enzyme as compared to those of the parent
amines.

In addition to the structural studies presented above, we
have also attempted to purify human brain PST to determine the
reaction mechanism of the enzyme and the true kinetic constants
for the substrate, dopamine. Initial attempts at purification
have proved somewhat difficult in that the transferase is
extremely unstable and cannot be dialyzed without substantial
loss of activity. The most promising method of purification to
date has been the application of chromatography with Affi-Gel
Blue. The elution pattern of a crude 100,000 x g supernatant
solution containing PST activity over an Affi-Gel Blue column is
illustrated in Figure 1. PST appears to be tightly bound to the

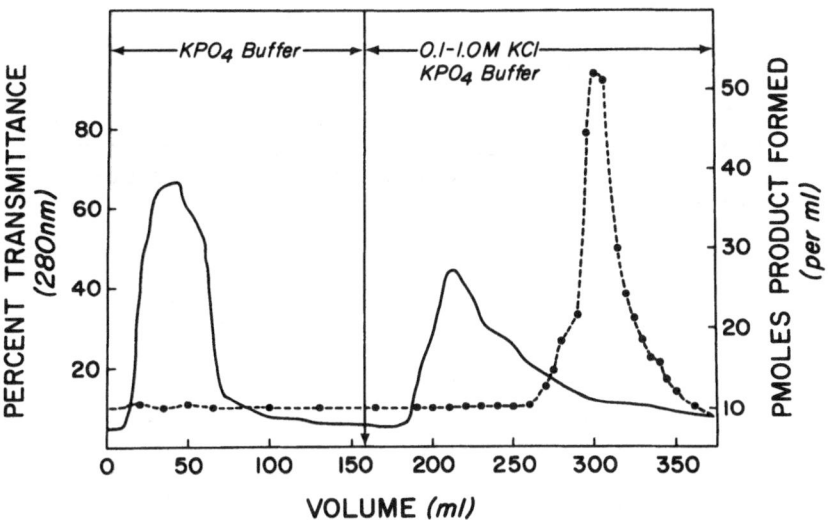

Figure 1. Elution pattern of PST from Affi-Gel Blue
Column at pH 7.4. Solid line represents protein, dashed
line represents PST activity

Gel and elutes at the tail end of the protein elution profile.
Although the elution pattern in Figure 1 suggests a rather high
degree of purification of PST, in reality, as the data in Table
2 indicate, only a five fold purification is achieved. Purification of PST is further hampered by the fact that on concen-

tration of the column eluate containing the enzyme, a decrease in specific activity occurs.

TABLE 2

Purification of human brain phenolsulfotransferase*

Enzyme Fraction	Units mg Prot.	Yield %	Purifi- cation
Crude homogenate	7.32	100.0	1.00
15,000 x G sup. sol.	-	-	-
100,000 x G sup. sol.	33.87	82.6	4.6
Affi-Gel Blue column (92-95)	217.9	18.9	29.8
Affi-Gel Blue column (91-99)	156.2	32.4	21.3
Concentrate of Affi- Gel Blue column fractions	103.7	11.8	14.2

*Total units of activity in the crude homogenate from 50 g of brain were 29,891 pmoles product formed/30 min.

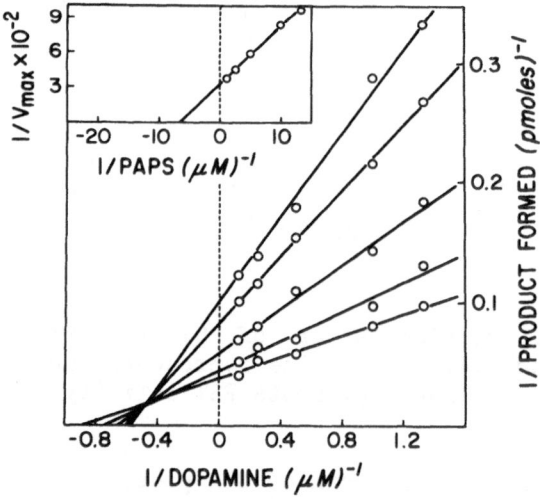

Figure 2. Lineweaver-Burk plot of sulfation of dopamine by partially purified human brain PST

Using the partially purified enzyme, the kinetic constants for dopamine and PAPS were calculated by multiple linear regression analysis of the data illustrated in the Lineweaver-Burke plots in Figure 2.

The Km value for dopamine is approximately 1.0 μM and that for PAPS is 0.16 μM. Prior studies have suggested that rat brain PST proceeds via an ordered reaction mechanism (Pennings et al., 1978). Assuming this kinetic model is also valid for the human enzyme then the Ks value for PAPS is calculated to be 0.63 μM.

Contribution of PST to Inactivation of Dopamine

The kinetic constant for dopamine calculated above is well within the range such that it potentially has biochemical significance in regard to the role of PST in the inactivation of this putative neurotransmitter. However, the relative contribution of this enzymatic reaction to the overall degradation of dopamine as compared to the other catecholamine inactivating reactions such as deamination and O-methylation cannot be assessed by measuring PST activity or its kinetic parameters alone. To evaluate this contribution of PST to the overall catabolic reactions, the relative activities and the kinetic constants of the three principle degradative enzymes, MAO, COMT and PST have to be quantitatively determined. As will be presented in the next section, our laboratory has been in the process of measuring the activities of MAO, COMT and PST in crude homogenates from frontal lobe of human brain. Based on the kinetic rate equations and constants calculated for each enzyme, estimates for the percent metabolism of dopamine by each of the three enzymatic pathways has been computed. The kinetic model for dopamine metabolism is based on the equations listed below.

$$(1) \quad v = \frac{-d[DA]}{dt} = \frac{(d[P]mao)a,b}{dt} + \frac{(d[P]comt)mb,sol.}{dt} + \frac{d[P]pst}{dt}$$

$$(2) \quad \frac{d[P]_{mao}}{dt} = \frac{V_{max}\,[DA]\,[O_2]}{K_m^{da}\,[O_2] + K_m^{O2}\,[DA] + [DA]\,[O_2]}$$

$$(3) \quad \frac{d[P]_{comt}}{dt} = \frac{V_{max}\,[DA]\,[SAM]}{K_m^{da}\,K_s^{sam} + K_m^{da}\,[SAM] + K_m^{sam}\,[DA] + [DA]\,[SAM]}$$

$$(4) \quad \frac{d[P]_{pst}}{dt} = \frac{V_{max}\,[DA]\,[PAPS]}{K_m^{da}\,K_s^{paps} + K_m^{da}\,[PAPS] + K_m^{paps}\,[DA] + [DA]\,[PAPS]}$$

Derivation of the equations above at varying concentrations of dopamine yields individual curves for the formation of product for each enzyme. There are two forms of MAO present in human brain that metabolize dopamine, type A and type B (Roth and Feor, 1978; White and Glassman, 1977). Studies in this laboratory (unpublished observations) have demonstrated that deamination of dopamine by both forms of MAO proceeds via a double-displacement reaction mechanism as indicated by the rate equation for MAO in #2 above. A recent publication from this laboratory (Roth, 1980) has revealed there are two distinct forms of COMT present in human brain, a soluble and a microsomal species. Preliminary studies are consistent with the reaction sequence for both forms of this transferase proceeding either by a random or an ordered mechanism and regardless, equation #3 above is compatible with both reaction sequences. The equation above for PST is based on preliminary findings which are consistent with this transferase proceeding via an ordered reaction mechanism with PAPS being the leading substrate. All studies that have been performed to date strongly imply that only a single species of PST is responsible for sulfation of the phenolic and catecholamines, including dopamine, in human brain (Renskers et al., 1980).

The kinetic constants for catabolism of dopamine by human brain MAO, COMT and PST are presented in Table 3. Of all the enzymes examined, PST has the highest affinity for dopamine

TABLE 3

Kinetic constants for dopamine catabolism by human brain*

Enzyme	Dopamine Km	Km	SAM Km	SAM Ks	PAPS Km	PAPS Ks
Type A MAO	107	85	--	--	--	--
Type B MAO	331	379	--	--	--	--
COMT-MB	3.3	--	3.1	2.1	--	--
COMT-sol.	274	--	N.C.	N.C.	--	--
PST	1.0	--	--	--	0.16	0.63

*Methods used are described in the text. Concentrations given are in μM. N.C. - not calculated

whereas MAO B has the lowest affinity. Although the data presented in this table may suggest that at low concentrations of dopamine PST becomes the principal inactivating enzyme, it is also dependent on the maximum velocity for each enzymatic reaction. Using the equations presented above, the Vmax for each enzyme was calculated by measuring the velocity at a given concentration of dopamine for each brain sample examined. As described in Methods, type A and B MAO activity was estimated by measuring dopamine deamination in the presence and absence of clorgyline. Total COMT activity was determined at 10 and 100 μM dopamine, and the Vmax value for microsomal and soluble COMT estimated by derivation of two simultaneous equations. The velocities and the derived Vmax values for dopamine metabolism for a typical sample of a crude homogenate preparation from frontal lobe of human brain are presented in Table 4. In contrast to the relative binding affinities, PST displayed the lowest activity whereas type B MAO had the highest activity.

TABLE 4

PST, MAO and COMT activities in human brain homogenate*

Enzyme	Dopamine (μM)	Velocity	Vmax
PST	10	0.0172	0.0228
Type A MAO	200	4.62	8.87
Type B MAO	200	4.99	21.8
COMT (Total)	10	0.161	--
	100	0.531	--
Soluble COMT	--	--	1.46
Membrane-bound COMT	--	--	0.146

*Activity is expressed as nmoles product formed/30 mins/mg protein. Assays were carried out in potassium phosphate buffer, pH 7.4. The concentrations of PAPS, oxygen and SAM were 1 μM, 218 μM and 85 μM, respectively.

Based on the data indicated in Tables 3 and 4, theoretical curves for the relative percent metabolism of dopamine by each enzyme were derived and the results are presented in Figure 3A. Saturating concentrations of the cofactors, S-adenosylmethionine (SAM) and PAPS, for COMT and PST, respectively, were used to calculate the enzyme activities. The oxygen concentrations used, 218 μM, is equivalent to the concentration of oxygen dissolved in water at 37°C. Under these conditions, MAO possessed the highest activity even at low concentrations of dopamine where the affinity of dopamine for PST as compared to that for MAO is almost 100 times greater. In this brain preparation sulfation of dopamine in the frontal lobe represents less than 10 percent of the total enzymatic activity for inactivation of dopamine. Although the relative contribution of the different enzymatic pathways in this brain preparation is representative of that seen in other samples, we have also noticed considerable varia-

82

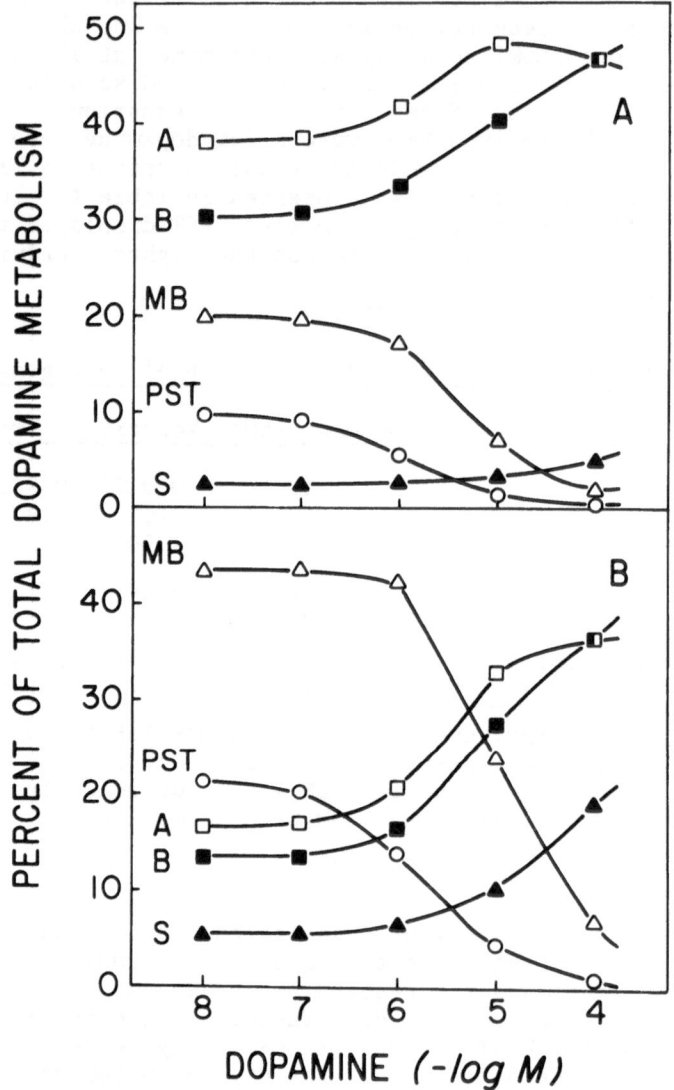

Figure 3A and B. Percentage contribution of MAO
type A and B, COMT (soluble and membrane bound) and PST
to metabolism of dopamine by a crude preparation of human
brain frontal lobe.

tion in the levels and specific activities of these enzymes in
the preparations examined. It should be emphasized here that
the data shown in Figure 3A does not take into account selective
localization of the enzymes involved in this catabolic process.
The relative contribution of PST activity in frontal lobe specimens may not be indicative of the activity in different areas or
cell types of human brain. If PST actually plays a major role
in dopamine metabolism, it can be anticipated that brain areas
rich in dopaminergic neurons may contain the highest levels of
PST.

The studies described above also afford us an opportunity
to estimate the relative importance of each enzymatic pathway
under varying conditions which may effect enzyme levels or
activities. For example, the relative percent metabolism via
each enzymatic pathway can be predicted in depressed patients
treated with a monoamine oxidase inhibitor drug. If it is
assumed that under these conditions both type A and B MAO
activity are inhibited approximately 80 percent, then the
results shown in Figure 3B are obtained. These data suggest
that at low concentrations of dopamine, membrane-bound COMT and
PST activities predominate and become the principle enzymes
involved in the destruction of dopamine. As the concentration
of dopamine increases and both these enzymes become saturated,
their relative contribution to the overall inactivation of
dopamine becomes greatly diminished.

CONCLUSIONS

Although the approach taken above does not take into
account many of the complicating factors associated with brain
heterogeneity, it does permit us to obtain a rough estimate of
the relative importance of PST activity in the human brain in
regard to its capacity to inactivate dopamine. The data clearly
demonstrate that even under the least likely conditions, PST
activity can account for approximately 10 percent of the total
enzymatic activity involved in the destruction of this catecholamine neurotransmitter. As techniques become more refined
better estimates of the relative contribution of PST to the
dopamine inactivating pathways in selected areas and cell types
of human brain will become available. Since PST is also capable
of conjugating norepinephrine and other structurally related
amines, similar studies can also be performed with these
biologically active agents.

Anderson, R.J. and Weinshilboum, R.M. (1980). Phenolsulfotransferase in Human Tissue: Radiochemical Enzymatic Assay and Biochemical Properties. Clin. Chim. Acta, 103, 79-90.

Bronaugh, R.L., Hattox, S.E., Hoehn, M.M., Murphy, R.C. and Rutledge, C.O. (1975). The Separation and Identification of Dopamine 3-0-Sulfate and Dopamine 4-0-Sulfate in Urine of Parkinsonian Patients. J. Pharmacol. Exp. Ther., 195, 441-452.

Foldes, A. and Meek, J.L. (1973). Rat Brain Phenolsulfotransferase - Partial Purification and Some Properties. Biochim. Biophys. Acta, 327, 365-374.

Goodall, M. and Alton, H. (1972). Metabolism of 3,4-Dihydroxyphenylalanine (L-DOPA) in Human Subjects. Biochem. Pharmacol., 21, 2401-2408.

Haggendal, J. (1963). The Presence of Conjugated Adrenaline and Noradrenaline in Human Blood Plasma. Acta Physiol. Scand., 59, 255-260.

Hart, R.F., Renskers, K.J., Nelson, E.B. and Roth, J.A. (1979). Localization and Characterization of Phenol Sulfotransferase in Human Platelets. Life Sci., 24, 125-130.

Jenner, W.N. and Rose, F.A. (1974). Dopamine-3-0-Sulfate, an End Product of L-DOPA Metabolism in Parkinson Patients. Nature, 252, 237-238.

Johnson, C.A., Baker, C.A. and Smith, R.T. (1980). Radioenzymatic Assay of Sulfate Conjugates of Catecholamines and DOPA in Plasma. Life Sci., 26, 1591-1598.

Kahone, Z., Essner, A.H., Kline, N.S. and Vestergaard, P. (1967). Estimation of Conjugated Epinephrine and Norepinephrine in Urine. J. Lab. Clin. Med., 69, 1042-1050.

Pennings, E.J.M., Vrielink, R. and VanKempen, G.M.J. (1978). Kinetics and Mechanism of the Rat Brain Phenol Sulfotransferase Reaction. Biochem. J., 173, 299-307.

Renskers, K.J., Feor, K.D. and Roth, J.A. (1980). Sulfation of Dopamine and Other Biogenic Amines by Human Brain Phenol Sulfotransferase. J. Neurochem., 34, 1362-1368.

Richter, D. (1940). The Inactivation of Adrenaline in Vivo in Man. J. Physiol., 98, 361-374.

Richter, D. and MacIntosh, F.C. (1941). Adrenaline Ester. Am. J. Physiol., 135, 1-5.

Roth, J.A. (1980). Presence of Membrane-Bound Catechol-O-Methyltransferase in Human Brain. Biochem. Pharmacol., 29, 3119-3122.

Roth, J.A. and Feor, K.D. (1978). Deamination of Dopamine and its 3-0-Methylated Derivative by Human Brain Monoamine Oxidase. Biochem. Pharmacol., 27, 1606-1608.

Rutledge, C.O. and Hoehn, M.M. (1973). Sulphate Conjugation and L-DOPA Treatment of Parkinsonian Patients. Nature, 244, 447-450.

Tyce, B.M., Sharpless, N.S., Kerr, F.W.L. and Muenter, M.D. (1980). Dopamine Conjugate in Cerebrospinal Fluid. J. Neurochem., 34, 210-212.

White, H.L. and Glassman, A.T. (1977). Multiple Binding Sites of Human Brain and Liver Monoamine Oxidase: Substrate Specificities, Selective Inhibitors, and Attempts to Separate Enzyme Forms. J. Neurochem., 29, 987-997.

Selective Inhibition of Sulfation in Vivo

Gerard J Mulder, Ina C M Halsema, Henk Koster, John H N Meerman
and K Sandy Pang

Department of Pharmacology, State University of Groningen, Groningen,
The Netherlands.
Department of Pharmaceutics, University of Houston,
Houston, Texas, USA

SULFATION AND COMPETING CONJUGATION REACTIONS: TOXIFI-
CATION AND DETOXICATION

Sulfation is one of the most versatile conjugation reactions which accepts phenolic and alcoholic hydroxyl
groups, hydroxamic acids and amines as acceptor groups
for conjugation. In this chapter only the sulfation of
low-molecular weight substances will be discussed;
most likely, completely unrelated sulfotransferases
are involved in the sulfation of macromolecules like
proteins and glycosaminoglycans, or phospholipids.
Further, there are no data as yet on the selective inhibition of sulfation of these types of substrates.
Both endogenous and xenobiotic substrates are metabolized by a number of sulfotransferases (Roy, 1981;
Jakoby et al., 1980); the resulting sulfate conjugates
are usually rapidly excreted in urine or bile. The toxicity of the parent compound is often lost after sulfate conjugation. Therefore, sulfation and other conjugation reactions have been considered to be detoxication reactions (Williams, 1947). However, in recent
years it has become clear that in some cases sulfate
conjugates are much more toxic than the parent compound. This is particularly important in the field of
chemical carcinogenesis, where it has been found that
sulfation of hydroxamic acids like N-hydroxy-2-acetyl-
aminofluorene leads to the generation of chemically
reactive, electrophilic ultimate carcinogens (Dè Baun
et al., 1970). The same applies to sulfation of 1'-
hydroxy saffrole at its secondary (allylic) alcohol
group (Wislocki et al., 1976). These and several other
examples have been reviewed recently (Mulder, 1981b).

Also in the field of steroid metabolism there is strong
evidence that many sulfate conjugates are not end-products of steroid catabolism, but are (obligatory?) intermediates in the interconversion of various steroids.
The specificity and kinetic parameters of some hydro-
xylases are remarkably different for some sulfated and
unconjugated steroids (Ingelman-Sundberg et al., 1975;
Gustafsson et al., 1976; Harvey & Hobkirk, 1977); the
sulfate conjugate may have a higher affinity. Further,
sulfate conjugates may behave pharmacokinetically completely different, as illustrated by the uptake and
metabolism of oestrone and oestrone sulfate respectively by the isolated perfused rat liver (Höller et al.,
1977).
For some steroid hormones the sulfate conjugate has
been suggested to be a storage form, that might be taken up specifically by tissues possessing the required
carrier, binding proteins and sulfatase needed to re-
activate the steroid hormone (Bernstein & Solomon,
1970; Dessypris, 1975; Lebeau & Beaulieu, 1973). In
contrast to this role of sulfation, the glucuronide
conjugates that usually are formed from the same substrates, are considered to be metabolic end-products,
that are eliminated in bile or urine.
The above shows that sulfation does not necessarily
mean detoxication or loss of biological actions. A
further complicating factor is that in many cases
other conjugation reactions compete with sulfation for
the same substrate. Especially glucuronidation with
its usually much higher capacity of conjugation,
though lower affinity, is an important competing reaction. As indicated above, glucuronides, in all cases
investigated so far, are much less toxic than their
sulfate counterparts, and are probably more rapidly
excreted because of their lower lipid-solubility than
sulfate conjugates. For catechols methylation by ca-
techol-0-methyltransferase (COMT) is an important competing pathway for conjugation, although also monophenolic drugs may be methylated (Guldberg & Marsden,
1975; Pazmino et al., 1979; Usdin et al., 1979). Some
methylated catechols have been implicated in mental
disorders (Baldessarini, 1975).
In order to study the role of sulfation in the metabolism and toxicity of various classes of its substrates,
it is almost imperative to have methods at one's dis-
posal that inhibit sulfation selectively and, if
possible, completely.

PENTACHLOROPHENOL (PCP) AND 2,6-DICHLORO-4-NITROPHENOL
(DCNP) AS SELECTIVE INHIBITORS OF SULFATION IN VIVO

Phenolic Acceptors
The selective inhibition of sulfation by pentachlorophenol (PCP) was found accidentally after the report
by Kobayashi et al. (1976) on the inhibition of sulfation of phenol by PCP in gold fish liver slices. When
the effect of PCP was tested in the rat in vivo on the
conjugation of harmol, a hallucinogenic drug (Figure 1),
PCP was found to inhibit highly effectively the sulfation of harmol (Mulder & Scholtens, 1977). Consequently,
the competing glucuronidation reaction was increased
because more of the substrate became available for
that conjugation (Table 1). By testing the conjugating

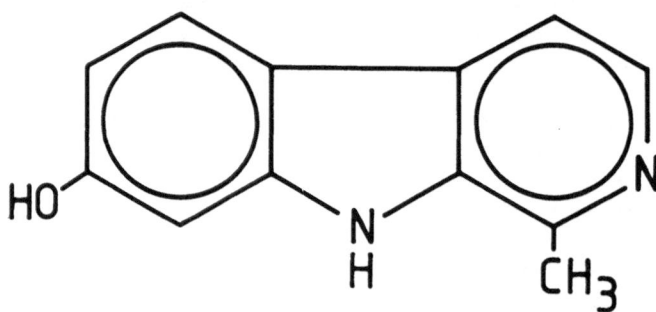

Figure 1. Harmol

enzyme activities in vitro, it was found that PCP and
DCNP inhibited the sulfotransferase but left UDP-glu-
curonosyltransferase unaffected. DCNP was slightly
more inhibitory than PCP; after one single dose of 26
μmol/kg DCNP the sulfation of harmol remained almost
completely inhibited for 24 hours. This makes it a
convenient drug for long-term inhibition of sulfation.
DCNP pretreatment did not affect the glucuronidation
of phenolphthalein, acetylation of procainamide ethobromide (PAEB) and glutathione conjugation of ethacry-
nic acid in the rat in vivo (Koster et al., 1979a).
Moreover, the biliary and urinary excretion of these
compounds, and of dibromsulphthalein (DBSP) were unaffected by the pretreatment. These findings confirmed
that the effect of DCNP in the rat was rather selective for sulfation.
The sulfation of several other phenolic substances was
similarly inhibited by DCNP, such as phenol (Koster et
al., 1979a), 4-nitrophenol and 1-naphthol (Mulder,
1981a). A number of other phenolic compounds inhibited

TABLE 1

The instantaneous inhibition of sulfation of
harmol in the rat in vivo by intravenous injection of 2,6-dichloro-4-nitrophenol

Interval between harmol and DCNP injection (min)	Harmol sulfate (μmol)	Harmol glucuronide (μmol)	n
0	0.08 ± 0.07	4.6 ± 0.3	4
5	1.87 ± 0.22	3.6 ± 0.2	4
10	2.28 ± 0.18	2.9 ± 0.1	4
15	2.68 ± 0.27	2.0 ± 0.4	4
180	5.32 ± 0.67	1.4 ± 0.3	6

Data are taken from Koster et al. (1981). DCNP
was first dissolved in a small volume of 0.1 N
NaOH and then diluted with 5% bovine serum albumin in saline; the final solution of pH 7 was
injected intravenously. DCNP was injected intravenously either at the same time as harmol,
or after the interval indicated in the Table.
The injections were in the femoral veins. The
dose of DCNP was 26 μmol/kg, and that of harmol 22 μmol/kg. Bile and urine of the anaesthesized rats were collected for three hours, and
harmol sulfate and harmol glucuronide were determined. The sum of the amounts of each conjugate excreted in bile and urine is given. The
mean ± SEM is given, and n is the number of rats
per group. The bodyweight of the rats was 300 g.

sulfation of harmol (Table 2), while several others
were not effective (Mulder & Scholtens, 1977; Koster
et al., 1979a, 1979b). For some of these phenols almost certainly the mechanism of inhibition is competitive inhibition for sulfation with harmol. However,
the exact biochemical mechanism of inhibition of sulfation by PCP and DCNP is unknown. Since the compounds
are also inhibitory in vitro, when PAPS, the co-substrate for sulfation, is added as such, a direct effect
at the level of the sulfotransferase is most likely.
No data are available on the metabolism of DCNP in the
rat. Presumably, the compound has a long half-life,
because of its long duration of action. The main
metabolite of PCP is 2,3,5,6-tetrachlorohydroquinone

TABLE 2

Inhibitory effect of some phenolic drugs on the sulfation of harmol in the rat in vivo

Pretreated with	n	Dose µmol/kg	Percentage of dose excreted as harmol sulfate
Control	8	–	22.1 ± 1.8
Pentachlorophenol	5	39	5.3 ± 0.9
2,6-Dimethyl-4-nitrophenol	4	26	22.2 ± 1.0
2,6-Dichloro-4-nitrophenol	4	26	2.2 ± 1.0
3-Methyl-2-nitrophenol	2	26	20.0
Salicylamide	2	26	22.1

Data are taken from Mulder and Scholtens (1977). The rats were pretreated with the compounds indicated in the Table by intraperitoneal injection, 45 minutes before the intravenous injection of harmol (20 µmol/kg). The kidneys of the anaesthesized rats were ligated, and bile was collected for 2 h after the injection of harmol. Harmol sulfate and harmol glucuronide were determined in the bile samples. The results are given as mean ± SEM, for n rats per group.

(excreted mainly as glucuronide conjugate), and about 10-15 percent of the dose was recovered as PCP glucuronide conjugate. No sulfate conjugate has been found (Ahlborg et al., 1974; Ahlborg & Thunberg, 1980). The mechanism of the inhibition may be the formation of a dead-end complex of the sulfotransferase with PCP or DCNP.
The reversibility of the inhibition of sulfation by DCNP has been tested in the isolated perfused rat liver (Koster et al., 1981). The sulfation of harmol was almost completely inhibited by a concentration of 60 µM DCNP in the perfusion medium, that was perfused single-pass through the liver (Pang & Gillette, 1978). It was rapidly reversible: when the drug was left out of the perfusion medium within a few minutes the sulfate conjugation returned to the control level or even slightly above (Figure 2).
This proves that the inhibition by DCNP is not irreversible inactivation of the sulfotransferase or the sulfate activating system in the liver.

DCNP is instantaneously effective when it is injected
intravenously in the rat as illustrated in Table 1.
Immediately upon injection the sulfation of harmol
stopped almost completely; thus, intravenous injection
of DCNP can be used to arrest sulfation of phenols in
vivo instantaneously.

Hydroxamic Acids

Initial experiments in vitro showed that PCP inhibited
the sulfation of N-hydroxy-phenacetin, a hydroxamic
acid, much more effectively than DCNP (Mulder & Scholtens, 1977). This was later confirmed with the well-
known carcinogen N-hydroxy-2-acetylaminofluorene (Meerman et al., 1980). Pretreatment with PCP and DCNP decreased the formation of the chemically reactive N-O-
sulfate ester of N-hydroxy-2-acetylaminofluorene and,
thereby, the formation of covalent adducts of this compound to macromolecules like protein, RNA and DNA. In
the recirculating isolated perfused rat liver the omission of inorganic sulfate from the perfusion medium resulted in the same decrease in the formation of [9-^{14}C]
-N-hydroxy-2-acetylaminofluorene adducts to protein,
RNA and DNA as the presence of PCP at normal levels of
inorganic sulfate. The effects of omission of sulfate
and addition of PCP were not additive, confirming that
both operated by the same mechanism, namely inhibition
of sulfation.
Interestingly, PCP and DCNP prevented the hepatotoxic
action of N-hydroxy-2-acetylaminofluorene in the rat
in vivo, suggesting that this is due to sulfation
(Meerman and Mulder, 1981). Thus, PCP seems to be a
useful tool to investigate the role that sulfation
plays in the carcinogenic action of N-hydroxy-2-acetyl-
aminofluorene and related compounds.

Steroids and Neurotransmittors

As yet no data are available on the effect of PCP and
DCNP on the sulfation of steroids or neurotransmittors.
It is to be expected that PCP and DCNP will easily
cross the blood-brain barrier.

TOXICITY OF PCP AND DCNP

Many data on the toxicity of PCP have been collected,
since the compound is widely used and, therefore, ubi-
quitously present as environmental contaminant. A big
difference in toxicity is observed between pure PCP
and commercial preparations due to the host of conta-
minations in industrially used PCP, such as many poly-

chlorinated aromatic compounds of high toxicity (Ahlborg & Thunberg, 1980). The toxicity of PCP and related chlorinated phenols has recently been reviewed
(Ahlborg & Thunberg, 1980).
One of the biochemically and toxicologically most important actions of PCP is its uncoupling action on oxidative phosphorylation that occurs at low concentrations in vitro and in isolated hepatocytes (Ahlborg &
Thunberg, 1980). Energy-dependent processes like the
uptake of BSP by isolated hepatocytes may, therefore,
be inhibited by PCP (Götz et al., 1980). It seems, however, that at doses effective in vivo in the preven-
tion of sulfation this action is of minor importance,
even though the elimination of PCP in the rat is slow,
the $t_{\frac{1}{2}}$ being approximately 20 hours (Braun et al.,
1977). The compound can be chronically administered
without signs of toxicity in the rat as long as the
level in the food does not exceed 500 ppm.
The fact that isolated rat hepatocytes showed the uncoupling action of PCP, whereas this is not obvious,
even at rather high doses, in vivo requires an explanation. An explanation may be that in vivo PCP is accumulated in certain tissues, whereas in the hepatocyte incubation no such 'sink' is available. More compounds
that have weak or no effect in vivo may be strongly
active in hepatocyte incubations: ethanol, salicylamide, phenolphthalein glucuronide and phenolphthalein
sulfate inhibit sulfation and glucuronidation of various compounds in isolated hepatocytes, whereas they
have hardly any effect in vivo, presumably because in
this case they are too rapidly excreted (Mulder, 1981a).
It seems therefore, that in spite of the toxic activities that PCP certainly possesses, it is a selective
inhibitor of sulfation at low doses in the rat in vivo.
For DCNP no data are available on its metabolism,
pharmacokinetic behaviour and toxicity.

SOME OTHER METHODS TO DECREASE SULFATION IN VIVO

Salicylamide and paracetamol (acetaminophen) have also
been employed to inhibit sulfation. In part such inhibition may have been due to a decreased availability
of sulfate as a result of sulfation of these drugs
themselves, as discussed in the chapter by Mulder and Krijgsheld.
Although salicylamide inhibits sulfation of harmol and
hydroxylated metabolites of benzpyrene (Andersson et
al., 1978; Burke et al., 1977) in hepatocytes, and
isoprenaline in the gut (George et al., 1974), yet

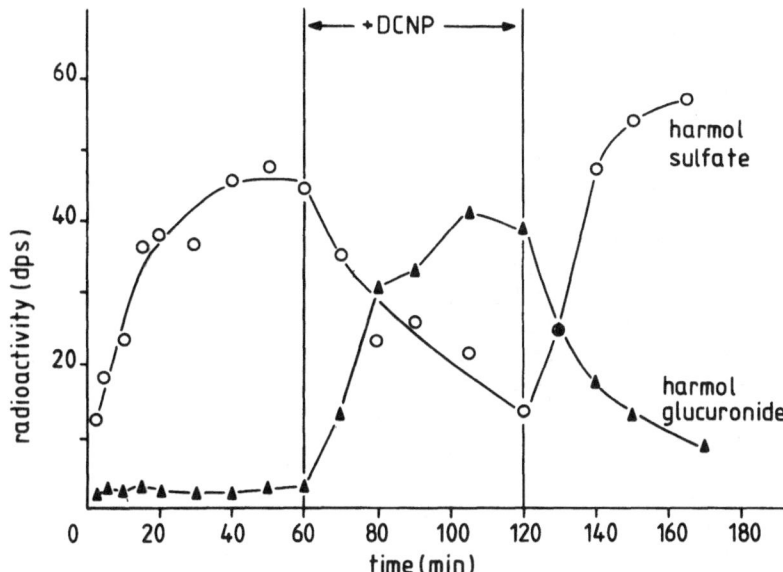

Figure 2. The inhibition of sulfation of
harmol by 2,6-dichloro-4-nitrophenol in
the single pass perfused rat liver.

The data are from Koster et al. (1981). The
liver was perfused with Krebs-Ringer buffer,
containing bovine serum albumin and dextran,
as described by Pang & Gillette (1978); in
addition 20% (v/v) washed sheep erythrocytes
were added. The liver flow was 10 ml per
minute. The perfusion medium contained a
tracer concentration of ^{3}H-(general) labeled
harmol. At t = 0 this was introduced in the
perfusion and the concentrations of radioactivity in harmol sulfate (0 - 0) and harmol
glucuronide (▲-▲) in the effluent are given
in the Figure. At t = 60 minutes a concentration of 60 μM 2,6-dichloro-4-nitrophenol was
added to the perfusion medium, and at t = 120
minutes it was left out again. The amount of
radioactivity excreted in bile during the perfusion had the following composition
From 0-60 min 94% sulfate and 6% glucuronide
                (20% of radioactivity in bile)
     60-120 "   43% sulfate and 57% glucuronide
                (33% of radioactivity in bile)
    120-180 "   57% sulfate and 43% glucuronide
                (26% of radioactivity in bile)

this drug is not very effective in vivo (Mulder & Scholtens, 1977). It is too rapidly eliminated after sulfation and glucuronidation (Levy & Matsuzawa, 1967). It has a very high affinity for phenolsulfotransferase (Davis, 1975). Paracetamol in vivo has the same 'problem' of its rapid elimination as glucuronide and sulfate conjugate. The dose of this drug must not be too high because of its glutathione-depleting effect and its hepatotoxicity (Jollow et al., 1974; Mitchell et al., 1975). Therefore, although paracetamol and salicylamide may have some usefulness in acute experiments, the analysis of the results will in general be too complicated to permit clear conclusions, because both glucuronidation and sulfation may be affected (Jollow et al., 1974; Mitchell et al., 1975; Mohan et al., 1976; Krijgsheld et al., 1981).
The same applies to diets low in sulfur-containing amino acids as discussed in the chapter of Mulder and Krijgsheld, which lead to adaptive changes in the conjugating enzyme activities and levels of the enzymes involved in for instance transsulfurylation (Mulder & Krijgsheld, 1981).

OUTLOOK

Neither PCP nor DCNP seem very attractive to use in human experiments and in that respect salicylamide and paracetamol are more useful. However, they are a starting point from which to look for other, possibly safer, inhibitors of sulfation. The advantage of PCP and DCNP as tools in experimental animals is the long duration of action, presumably related to their slow elimination, and their apparent selectivity. Their mechanism of action, however, still awaits elucidation.

REFERENCES

Ahlborg, U.G. and Thunberg, T.M. (1980). Chlorinated phenols: occurrence, toxicity, metabolism and environmental impact. CRC Crit. Revs Toxicol. 7, 1-36.

Ahlborg, U.G., Lindgren, J.E. and Mercier, M. (1974). Metabolism of pentachlorophenol. Arch. Toxicol. 32, 271-281.

Andersson, B., Berggren, M. and Moldeus, P. (1978). Conjugation of various drugs in isolated hepatocytes. Drug. Metabol. Disposit. 6, 611-616.

Baldessarini, R.J. (1975). Biological transmethylation involving S-adenosylmethionine: implications for neuropsychiatry. Int. Rev. Neurobiol. 18, 41-67.

Bernstein, S. and Solomon, S (eds) (1970). Chemical and 95
Biological Aspects of Steroid Conjugation. Springer
Verlag, Berlin, Germany.
Braun, W.H., Young, J.D., Blau, G.E. and Gehring, P.J.
(1977). The pharmacokinetics and metabolism of pentachlorophenol in rats. Toxicol. appl. Pharmacol. 41,
395-406.
Burke, M.D., Vadi, H., Jernström, B. and Orrenius, S.
(1977). Metabolism of benzo(a)pyrene with isolated
hepatocytes and the formation and degradation of
DNA-binding derivatives. J. Biol. Chem. 252, 6424-
6431.
Davis, D.C. (1975). Radioisotopic assay for rat liver
sulfotransferase activity. Biochem. Pharmacol. 24,
975-978.
DeBaun, J.R., Miller, E.C. and Miller, J.A. (1970). N-
Hydroxy-2-acetylaminofluorene sulfotransferase. Cancer Res. 30, 577-595.
Dessypris, A.G. (1975). Testosterone sulfate, its biosynthesis, metabolism, measurement, function and
properties. J. Steroid Biocher.. 6, 1287-1292.
George, C.F., Blackwell, E.W. and Davies, D.S. (1974).
Metabolism of isoprenaline in the intestine. J.
Pharm. Pharmacol. 26, 265-267.
Götz, R., Schwarz, L.R. and Greim, H. (1980). Effects
of pentachlorophenol and 2,4,6-trichlorophenol on
the disposition of sulfobromophthalein and respira-
tion of isolated liver cells. Arch. Toxicol. 44,
147-155.
Guldberg, H.C. and Marsden, C.A. (1975). Catechol-O-
methyltransferase: pharmacological aspects and physiological role. Pharmacol. Revs. 27, 135-206.
Gustafsson, J.A. and Ingelman-Sundberg, M. (1976). Multiple forms of cytochrome P-450 in rat liver microsomes. Eur. J. Biochem. 64, 35-43.
Harvey, P.R.C. and Hobkirk, R. (1977). The metabolism
of estrone and estradiol and their 3-sulfates by
female guinea pig liver microsomes. Steroids 30,
115-128.
Höller, M., Grochtman, W., Napp, M. and Breuer, H.
(1977). Studies on the metabolism of oestrone sulphate. Biochem. J. 166, 363-371.
Ingelman-Sundberg, M., Rane, A. and Gustafsson, J.A.
(1975). Properties of hydroxylase systems in the human fetal liver active on free and sulfoconjugated
steroids. Biochemistry 14, 429-432.
Jakoby, W.B., Sekura, R.D., Lyon, E.S., Marcus, C.J.
and Wang, J.L. (1980). Sulfotransferases, in Metabolic Basis of Detoxication (eds. W.B. Jakoby, J.R.
Bend and J. Caldwell), Vol. 2, chapter 11, Academic Press, New York, N.Y.

96   Jollow, D.J., Thorgeirsson, S.S., Potter, W.Z., Hashi-
        moto, M. and Mitchell, J.R. (1974). Acetaminophen-
        induced hepatic necrosis VI. Pharmacology 12, 251-
        271.
      Kobayashi, K., Kimura, S. and Akitake, H. (1976). Stu-
        dies on the metabolism of chlorophenols in fish, VII
        Bull. Jap. Soc. Sci. Fish. 42, 171-177.
      Koster, H., Scholtens, E. and Mulder, G.J.(1979a). In-
        hibition of sulfation of phenols in vivo by 2,6-di-
        chloro-4-nitrophenol. Med. Biol. 57, 340-344.
      Koster, H., Scholtens, E. and Mulder, G.J. (1979b).
        Inhibition of sulphation of phenolic substances by
        the carboxylase inhibitor bis-(p-nitrophenyl)-phos-
        phate in the rat in vivo. Biochem. Pharmacol. 28,
        2685-2686.
      Koster, H., Halsema, I.C.M., Pang, K.S. and Mulder, G.
        J. (1981). Inhibition of sulfation of harmol in the
        perfused rat liver by 2,6-dichloro-4-nitrophenol.
        Submitted.
      Krijgsheld, K.R., Scholtens, E. and Mulder, G.J. (1981)
        An evaluation of methods to decrease the availabili-
        ty of inorganic sulphate for sulphate conjugation in
        the rat in vivo. Submitted.
      Lebeau, M. and Beaulieu, E. (1973). On the significance
        of the metabolism of steroid hormone conjugates. In
        Metabolic Conjugation and Metabolic Hydrolysis (ed.
        W.H. Fishman), vol. 3, p. 151-187, Academic Press,
        New York, N.Y.
      Levy, G. and Matsuzawa, T. (1967). Pharmacokinetics of
        salicylamide elimination in man. J. Pharmacol. exp.
        Therap. 156, 285-293.
      Meerman, J.H.N., Van Doorn, A.B.D. and Mulder, G.J.
        (1980). Inhibition of sulfate conjugation of N-
        hydroxy-2-acetylaminofluorene in isolated perfused
        rat liver and in the rat in vivo by pentachlorophe-
        nol and low sulfate. Cancer Res. 40, 3772-3779.
      Meerman, J.H.N. and Mulder, G.J. (1981). Prevention of
        hepatotoxic action of N-hydroxy-2-acetylaminofluo-
        rene in the rat by inhibition of N-O-sulfation by
        pentachlorophenol. Submitted.
      Mitchell, J.R., Potter, W.Z., Hinson, J.A., Snodgrass,
        W.R., Timbrell, J.A. and Gillette, J.R. (1975).
        Toxic Drug Reactions. In Handbook of exp. Pharmacol.
        vol. 28, part 3, p.383-419. Springer Verlag, Berlin,
        Germany.
      Mohan, L.C., Grantham, P.H., Weisburger, E.K., and
        Weisburger, J.H. (1976). Mechanisms of the inhibito-
        ry action of p-hydroxyacetanilide on carcinogenesis
        by N-2-fluorenylacetamide or N-hydroxy-N-2-fluore-
        nylacetamide. J. Natl. Cancer Inst. 56, 763-768.

Mulder, G.J. (1981a). Sulfation in vivo and in isolated intact cell preparations. In Sulfation of Drugs and Related Compounds (ed. G.J. Mulder), chapter 6, CRC Press, Boca Raton, FL.

Mulder, G.J. (1981b). Generation of reactive intermediates from xenobiotics by sulfate conjugation. In Sulfation of Drugs and Related Compounds (ed. G.J. Mulder), chapter 8, CRC Press, Boca Raton, FL.

Mulder, G.J. and Krijgsheld, K.R.(1981). The availability of cofactors for conjugation as rate limiting step for conjugation in vivo. In Nutrition and Drug Metabolism/Disposition (eds. T.C. Campbell and D.A. Roe), Marcel Dekker, New York, N.Y.

Mulder, G.J. and Scholtens, E. (1977). Phenolsulphotransferase and UDP glucuronyltransferase from rat liver in vivo and in vitro. 2,6-Dichloro-4-nitrophenol as selective inhibitor of sulphation. Biochem. J. 165, 553-559.

Pang, K.S. and Gillette, J.R. (1978). Kinetics of metabolite formation and elimination in the perfused rat liver preparation. J. Pharmacol. exp. Therap. 207, 178-194.

Pazmino, P., Rogoff, F. and Weinshilboum, R. (1979). Inhibition of human erythrocyte phenol-O-methyltransferase in uremia. Clin Pharmacol. Therap.26, 464-472.

Roy, A.B. (1981). Sulfotransferases. In Sulfation of Drugs and Related Compounds (ed. G.J. Mulder), chapter 5, CRC Press, Boca Raton, FL.

Usdin, E., Borchardt, R.T. and Creveling, C.R. (eds) (1979). Transmethylation. Elsevier Biomedical Press, Amsterdam.

Williams, R.T. (1947). Detoxication Mechanisms. Chapman and Hall, London.

Wislocki, P.G., Borchert, P., Miller, J.A. and Miller, E.C. (1976). The metabolic activation of the carcinogen 1'-hydroxysafrole in vivo and in vitro and the electrophilic reactivities of possible ultimate carcinogens. Cancer Res. 36, 1686-1695.

Phenolsulphotransferase in Human Tissue: Evidence for Multiple Forms

Glen Rein, Vivette Glover and Merton Sandler

Bernhard Baron Memorial Research Laboratories and Institute of Obstetrics and Gynaecology, Queen Charlotte's Maternity Hospital, Goldhawk Road, London W6 0XG, UK

ABSTRACT

We have examined human platelet phenolsulphotransferase (PST) with a wide range of different substrates
and compared its activity pattern with that of enzyme
in other sites in the human body. The main finding was
that the dopamine/phenol activity ratio varied considerably from tissue to tissue, ranging from 5.6 in the
jejunum to 0.18 in the adrenal, pointing to the presence of more than one enzyme. Further evidence for the
presence of two forms of PST has been obtained by the
use of the inhibitor, dichloronitrophenol (DCNP), which
selectively inhibited phenol- compared with dopamine-
conjugating activity in both platelet and brain. We
therefore propose that the two forms of the enzyme be
denoted PST P (for phenol) and PST M (for monoamine
and monoamine metabolites) respectively. The specific
activity of the enzymes towards both phenol and dopamine varied considerably in the different human tissues
examined, both being most active in the jejunum and
least in the brain. The high activity in the intestine
supports the view that the enzymes are important in
dealing with exogenous phenols. The comparatively low
activity in the brain raises the question of whether
sulphoconjugation is functionally significant in this
tissue.

INTRODUCTION

In 1876 phenyl sulphate was first isolated in urine
from dogs fed phenol (Baumann,1876). Since that time,
a wide variety of phenols, steroids, bile acids, proteo-

glycans and cerebrosides have been shown to be excreted as sulphate conjugates in mammalian urine (Dodgson, 1977). Some of the many sulphotransferases involved in such conjugation reactions have been studied and appear to be active toward a broad range of physiological and artificial substrates.

It is therefore likely that several enzymes with different degrees of specificity are responsible. Early studies by Bostrom and Wengle (1967) examined the ability of different human tissues to conjugate a variety of steroid and phenolic substrates and provided evidence to indicate that different enzymes are responsible for steroid and phenol sulphoconjugation. Subsequent attempts to separate the sulphate-conjugating enzyme, phenolsulphotransferase (PST) physically from steroid sulphotransferase have met with varying degrees of success. Even so, partially-purified preparations of steroid sulphotransferase have been obtained which lack activity toward phenols, confirming that steroids and phenols can be sulphoconjugated by separate enzymes (Banerjee and Roy,1966; Adams and Chulavatnatol,1967;Adams and Poulos,1967). There is also evidence of heterogeneity among the steroid sulphotransferases (Nose and Lipmann,1958; Holcenberg and Rosen,1965; Banerjee and Roy,1967). Whether this enzyme group plays a significant role in the sulphoconjugation of phenols is unknown.

Less information is available about PST than about steroid sulphotransferases. PST has traditionally been assayed in rat or guinea pig tissues *in vitro* using different non-physiological phenols and naphthalenes. A wide variety of biologically active monoamines and their metabolites may be conjugated both, *in vitro* and *in vivo*, in the rat (Foldes and Meek,1973; Meek and Neff,1973). However, there are substantial differences in the conjugate excretion pattern of man and rat (Caldwell,1976,1980) so that rat PST should be viewed with caution as a model for the human enzyme. The recent discovery of PST activity in human platelets, able to metabolize monoamines and their metabolites (Hart et al.,1979; Rein et al.,1981a) has opened the way to clinical studies. Activity towards these physiological substrates has also been found in human brain (Renskers et al.,1980), erythrocytes (Anderson and Weinshilboum,1979), kidney (Anderson and Weinshilboum,1980) and jejunum (Anderson and Weinshilboum,1980).

We describe here the substrate specificity of PST in platelets and other human tissues using phenol and a wide variety of monoamines and their metabolites. By characterizing the enzyme in different sites, we hoped to determine whether platelet activity could be used as a yardstick for enzyme activity elsewhere in the body, particularly the brain. If more than one enzyme were present, one would expect the ratio of activity values with different substrates to vary from tissue to tissue. If several substrates were metabolized by the same enzyme, their activity ratio would remain constant, independent of site of tissue origin. This approach has previously been used to distinguish the multiple forms of monoamine oxidase (Tipton et al.,1976) and benzylamine oxidase (Lewinsohn et al., 1980) but has not been applied to the problem of identifying more than one form of PST.

METHODOLOGY

PST activity was routinely assayed using the radiometric method of Foldes and Meek (1973) using 0.4 μM (^{35}S) 3'-phosphoadenosine-5'-phosphosulphate (PAPS) as sulphate donor at pH 7.2. Diluted tissue homogenates were prepared from fresh or biopsy material in all cases, except brain which was obtained postmortem. Before characterizing the enzyme in different tissues, we examined the linearity of the reaction with increasing tissue concentration. A marked lack of linearity was observed for all tissue homogenates at higher concentrations but was most apparent with adrenal gland and jenunum (Rein et al.,1981c) (Figure 1). Care was accordingly taken to use concentrations from the linear range for each tissue in subsequent studies. This should have minimized the effects of any possible endogenous PST inhibitors (Anderson and Weinshilboum,1979).

PLATELET PST

The platelet enzyme catalyzes the sulphoconjugation of a wide range of monoamines and their metabolites (Rein et al.,1981a). Table 1 shows the activity obtained with a range of substrates at 30 μM. In general, the amines were more actively metabolized than the alcoholic metabolites which, in turn, were better substrates than the acidic metabolites. Sulphotransferase activity towards these acidic metabolites has not previously been reported for human tissues. At 30 μM, low activity towards 4-hydroxy-3-methoxymandelic acid (VMA) was observed with no detectable activity toward homovanillic acid (HVA) and 3,4-dihydroxyphenylacetic acid (DOPAC). Only at the high concentration of 3 mM was substantial activity obtained with VMA and HVA, although DOPAC-conju−

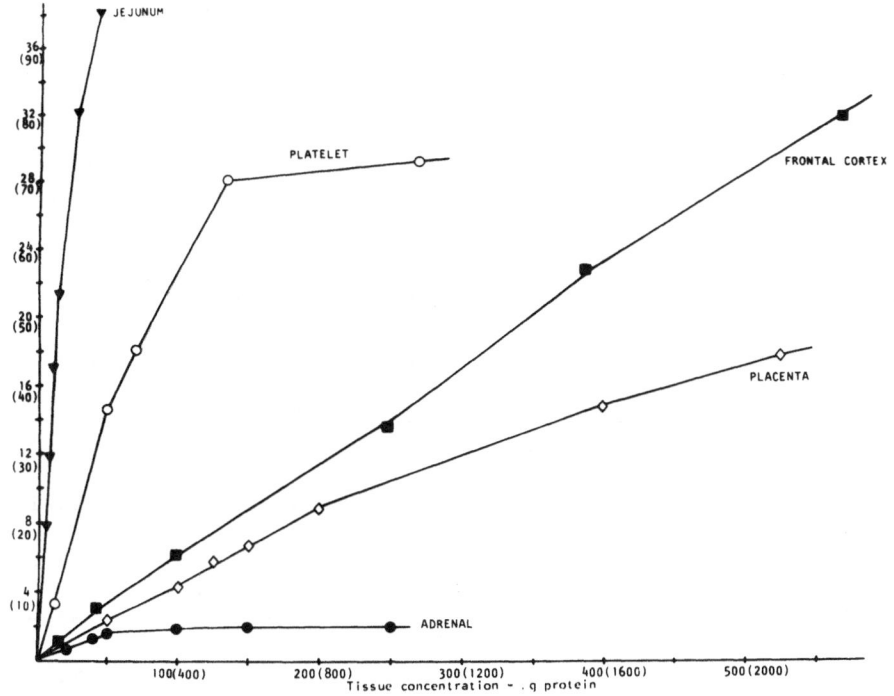

Figure 1

Effect of increasing tissue concentration on PST activity towards
30 μM dopamine or tyramine. Diluted pooled homogenates were used
in all cases except brain where a 30,000 g supernatant was used.
All tissues were preincubated for 10 min with 1 mM pargyline
except the frontal cortex and were assayed in its presence. The
ordinate represents amount of sulphate conjugate formed expressed
as c.p.m. x 10^3. The numbers along axes in parentheses apply to
the frontal cortex. Each point is the mean of duplicate determinations (from Rein et al.,1981c).

gating activity was still absent. This finding is surprising since substantial amounts of conjugated DOPAC
are excreted in normal human urine (von Euler et al.,
1959; Goodall and Alton,1968; Smith and Weil-Malherbe,
1971). If this conjugate is confirmed to be predominantly sulphate in nature, the possibility arises that
a distinct PST for DOPAC is localized in certain peripheral tissues. It has been noted that as much as 75%
of acid sulphates are lost during barium sulphate precipitation to remove unreacted PAPS (Foldes and Meek,
1973). However, this explanation seems unlikely to
account entirely for the observed lack of activity since the same assay procedure yields substantial DOPAC-
conjugating activity in rat tissues (Foldes and Meek,
1973; Meek and Neff,1973).

TABLE 1. Ability of human platelets to sulphoconjugate a variety of biologically active monoamines, precursors and metabolites

Substrate	Concentration	Specific activity	Relative activity
m-Tyramine	30 μmol/l	25.6 ± 2.5 (4)	116
Normetadrenaline	30 μmol/l	24.7 ± 1.6 (3)	112
Metadrenaline	30 μmol/l	24.2 ± 1.7 (3)	109
Dopamine	30 μmol/l	22.1 ± 1.2 (9)	100
4-Hydroxy-3-methoxy-phenylethanol (HMPE)	30 μmol/l	21.9 ± 2.4 (3)	99.1
p-Tyramine	30 μmol/l	20.7 ± 0.6 (5)	93.6
Isoprenaline	30 μmol/l	19.2 ± 1.8 (4)	86.9
m-Octopamine	30 μmol/l	18.0 ± 1.3 (5)	81.4
Noradrenaline	30 μmol/l	13.9 ± 0.7 (7)	62.9
Adrenaline	30 μmol/l	13.9 ± 0.7 (7)	62.9
3-Methoxytyramine (3MT)	30 μmol/l	13.0 ± 2.4 (5)	58.8
Phenol	30 μmol/l	8.9 ± 1.1 (5)	40.3
p-Octopamine	30 μmol/l	7.3 ± 0.5 (4)	33.0
4-Hydroxy-3-methoxy-phenylglycol (HMPG)	30 μmol/l	4.5 ± 0.4 (9)	20.4
3,4-Dihydroxyphenyl-glycol (DHPG)	30 μmol/l	3.3 ± 0.8 (4)	14.9
5-Hydroxytryptamine	30 μmol/l	3.0 ± 0.3 (6)	13.6
4-Hydroxy-3-methoxy-mandelic acid (VMA)	30 μmol/l	1.45 ± 0.2 (6)	6.6
	3 mmol/l	14.3 ± 0.6 (3)	64.7
Homovanillic acid (HVA)	30 μmol/l	N.D.	-
	3 mmol/l	8.3 ± 2.6 (3)	37.6

TABLE 1 (continued)

Substrate	Concentration	Specific activity	Relative activity
3,4-Dihydroxyphenyl-acetic acid (DOPAC)	30 μmol/l 3 mmol/l	N.D. N.D.	-
3,4-Dihydroxyman-delic acid (DOMA)	30 μmol/l	N.D.	-
3,4-Dihydroxyphenyl-alanine (DOPA)	30 μmol/l	N.D.	-
Tyrosine	30 μmol/l	N.D.	-
5-Hydroxyindole-acetic acid	30 μmol/l	N.D.	-
Phenylethylamine	30 μmol/l	N.D.	-

PST activity was assayed in a pooled unwashed diluted platelet homogenate from 4 healthy volunteers.
PST activity (mean ± SEM) is expressed as pmoles product formed/ min/mg protein (specific activity) or as percentage of activity obtained with dopamine, set at 100% (relative activity). The number of independent experiments is given in parentheses using different pooled platelet preparations from 4 subjects. N.D. = no detectable activity below 0.9 pmoles product formed/min/mg protein (from Rein et al.,1981a).

Platelet PST was also found to be incapable of forming sulphate conjugates of L-3,4-dihydroxyphenylalanine, 5-hydroxyindoleacetic acid, L-tyrosine and phenylethylamine (Rein et al.,1981a). The absence of activity toward tyrosine *in vitro* has previously been noted in rat tissues, despite the urinary excretion of tyrosine sulphate in mammalian urine (Hext et al.,1973). The lack of activity toward phenylethylamine, which differs from tyramine only in the lack of a ring hydroxyl group, supports the concept that PST catalyzes the formation of O-sulphate conjugates.

The structure-activity relationships arising from the pattern of substrate activities is presented in Table 2 in greater detail. Methylation of a ring hydroxyl group increases the activity of PST toward a substrate if a β-hydroxyl group is also present on the side-chain; thus the enzyme is more active towards

TABLE 2. Structure-activity relationships for a number of biol-
ogically-active monoamines and their metabolites

(A) R_2, R_1 Amines — $-C-C-N-R_4$ with R_3

(B) R_2, R_1 Alcohols — $-C-C-R_4$ with R_3

(C) R_2, R_1 Acids — $-C-COOH$ with R_3

	R_1	R_2	R_3	R_4	Relative activity (Dopamine = 100)
(A)					
m-Tyramine	H	OH	H	H	116
Normetadrenaline	OH	OCH_3	OH	H	112
Metadrenaline	OH	OCH_3	OH	CH_3	109
m-Octopamine	H	OH	OH	H	81.4
Isoprenaline	OH	OH	OH	$(CH_3)3$	86.9
Dopamine	OH	OH	H	H	100
p-Tyramine	OH	H	H	H	93.6
Noradrenaline	OH	OH	OH	H	62.9
Adrenaline	OH	OH	OH	CH_3	62.9
3-Methoxytyramine	OH	OCH_3	H	H	58.8
Phenol	OH	H	H	H	40.3
p-Octopamine	OH	H	OH	H	33.0
(B)					
HMPE	OH	OCH_3	H	OH	99.1
HMPG	OH	OCH_3	OH	OH	20.4
DHPG	OH	OH	OH	OH	14.9
(C)					
VMA	OH	OCH_3	OH	–	6.6
HVA	OH	OCH_3	H	–	N.D.
DOPAC	OH	OH	H	–	N.D.

TABLE 2 (continued)   105

Relative activity data from Table 1, obtained at 30 µM for all
substrates, are used. For abbreviations, see Table 1.

normetadrenaline and metadrenaline than their respective β-hydroxylated counterparts, noradrenaline and adrenaline. However, ring methylation results in decreased activity if the β-hydroxyl group is absent, 3-meth-
oxytyramine (3MT) being less actively metabolized than
dopamine. The presence of a β-hydroxyl group, as in noradrenaline and 4-hydroxy-3-methoxyphenylglycol (HMPG),
results in decreased sulphating activity as compared
with dopamine and 4-hydroxy-3-methoxyphenylethanol
(HMPE). m-Tyramine and m-octopamine are more actively
metabolized than their corresponding p-hydroxylated
compounds. The presence of a methylated amino group
does not change the activity of adrenaline relative to
noradrenaline. The alcoholic metabolites of noradrenaline, HMPG and 3,4-dihydroxyphenylglycol, but not of dopamine, HMPE, are less actively metabolized than their
corresponding amines.

KINETICS OF PLATELET PST

Kinetic analyses of platelet PST activity (Rein et al.,
1981a) are presented in Table 3. Apparent K_m values for
selected substrates varied over a 10,000-fold range
with the lowest K_m value of 0.3 µM being obtained for
3MT and the highest, of 3700 µM, being observed with VMA.
These large differences explain the relatively similar
V_{max} values which give a very different order of substrate activity from that based on specific activity at
30 µM. The platelet enzyme has a relatively high affinity for the catecholamines, which all possess apparent
K_m values at or below 5 µM, and a relatively low affinity for the acid metabolite, VMA, with its apparent K_m
value of 3.7 mM. These results suggest that, at low substrate concentrations, the catecholamines are preferentially metabolized. However, the acid metabolites are
probably present physiologically in much higher concentrations since microgram quantities only of the catecholamines are excreted in human urine whereas their
acid metabolites are present in milligram concentrations (Ruthven and Sandler,1969).

Lineweaver-Burk plots also revealed that substrate
inhibition occurred with all substrates analyzed. Substrate inhibition varied over a thousand-fold range with
inhibition occurring above 2 µM for 3MT and above 3 mM

TABLE 3. Kinetic parameters for human platelet PST

Substrate	K_m μM	V_{max} pmoles/min/ mg protein	Substrate inhibition μM
Dopamine	3.0	31.2	> 25
p-Tyramine	91.0	35.8	> 300
3-Methoxytyramine	0.30	37.2	> 2.0
Noradrenaline	5.0	24.4	> 10
Adrenaline	2.7	23.8	> 40
5-Hydroxytryptamine	159	13.5	> 500
HMPG	435	39.8	>1000
VMA	3700	13.6	>3000
Phenol	6.7	20.6	> 400

Apparent Michaelis constant (K_m) and maximum velocity (V_{max}) for the PST reaction were calculated from Lineweaver-Burk plots obtained from one experiment, using a pooled platelet preparation from 4 normal adult volunteers. Values obtained during two or three subsequent experiments did not vary more than 20% from those reported here (Rein et al.,1981a).

for VMA. Substrate inhibition was, therefore, apparent at subsaturating concentrations for some substrates or at concentrations of ten times the apparent K_m value for other substrates. In general, substrates with low apparent K_m values showed the lowest concentration requirements for substrate inhibition. The substrate inhibition pattern confirms suggestions made by previous authors (Hart et al.,1979; Renskers et al.,1980) to account for the different substrate specificity patterns obtained at different substrate concentrations.

The substrate specificity pattern of human platelets is substantially different from that observed in tissues from other species. In rat brain and liver, for example, relatively low PST activity towards the amines has been observed at pH 7.8 (Meek and Neff,1973) and the acidic metabolites were also relatively more active in these tissues. The pH optimum of 9 for dopamine and HMPG (Meek and Neff,1973) is in contrast with human tissue PST which has a pH optimum of approximately 6.5 for HMPG (Anderson and Weinshilboum,1980). These results indicate that certain properties of rat PST are substantially different from those of the human enzyme. Taken together with the different sulphate conjugate excretion pattern of the two species (Caldwell,1976,1980), they reinforce the impression that the rat is not a good model for studying sulphoconjugation in man.

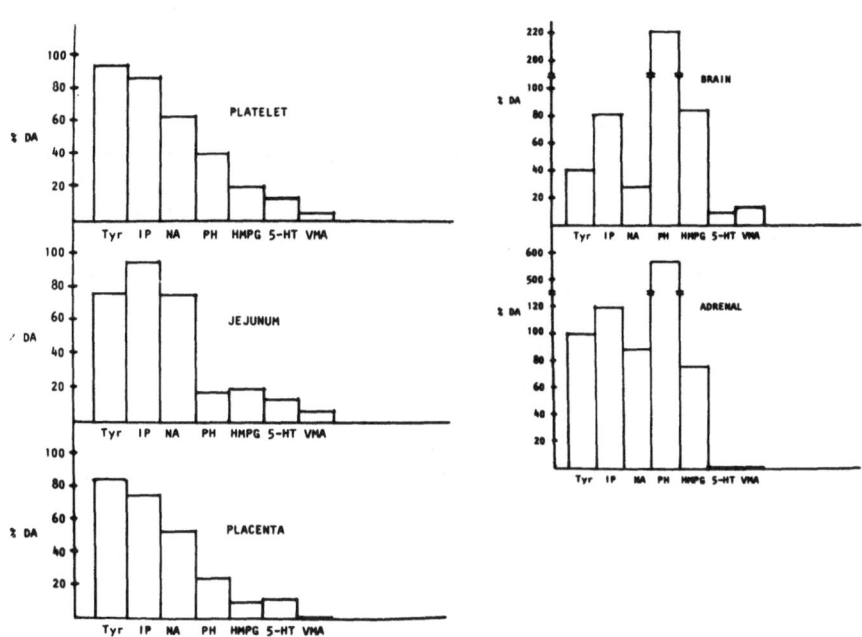

Figure 2

Substrate specificity pattern of PST in different human tissues. Activity was assayed in diluted pooled tissue homogenate using 1 mM pargyline and 30 µM of each substrate. Pooled homogenates from 4 subjects were used for the platelet and jejunum preparations and from 3 subjects for the placental preparation. The frontal cortex was assayed with 8 mM dithiothreitol, and without pargyline, using a 30,000 g supernatant from 3 subjects. Results are expressed relative to the activity obtained with 30 µM dopamine (= 100) in the same experiment. Each value represents the mean of 4 determinations (from Rein et al.,1981c).

Tyr = tyramine, IP = isoprenaline, NA = noradrenaline, PH = phenol, HMPG = 4-hydroxy-3-methoxyphenylglycol, 5-HT = 5-hydroxytryptamine, VMA = 4-hydroxy-3-methoxymandelic acid, DA = dopamine.

SUBSTRATE SPECIFICITY OF PST IN OTHER HUMAN TISSUES
Fig.2 shows the substrate specificity of PST in several
other human tissues, expressed as a percentage of activity with dopamine (Rein et al.,1981c). The patterns in
platelet, jejunum and placenta were very similar, with
dopamine being the most active substrate followed in decreasing order by tyramine, isoprenaline, noradrenaline,
phenol, HMPG, 5-hydroxytryptamine (5-HT) and VMA. However, the substrate specificity patterns for adrenal
gland and frontal cortex were quite different. Compared
with dopamine, each showed relatively more activity toward
phenol and HMPG than the other tissues. A relatively increased activity toward these substrates was also observed in the neural crest tumours, phaeochromocytoma and
neuroblastoma (Rein and Sandler,1981), both of which may
be derived from the adrenal medulla (Fig.3). These results suggest that in brain and adrenal tissues, phenol
and HMPG are metabolized by a different enzyme. This
point is brought out further in Table 4 where it will be
seen that the PST activity ratio between dopamine and
phenol was approximately the same in gut, platelets and
placenta but was reduced by a factor of 30 in the adrenal and in adrenal tumours (Rein et al.,1981b). The ratio
in the brain was also reduced although only by a factor
of 4. A similar pattern was observed for the dopamine to
HMPG ratio except that the brain had as low a ratio as
the adrenal tissues. In contrast to these values, the
activity ratio for dopamine and tyramine was constant in
the different tissues examined. These results then suggest that at least two enzymes are involved in the sulphoconjugation of phenol and the monoamines in human
tissues.

Because of the relatively low activity of PST toward
the acid metabolites at 30 µM, it was of interest to determine whether a similar substrate specificity pattern
would be maintained in the different tissues at 3 mM
concentrations. The relative activity data in Table 5
indicate that similar activity towards VMA, HVA and
DOPAC was present in platelet, gut and placenta (Rein et
al.,1981c). Like the other tissues, the adrenal showed
high activity towards VMA relative to dopamine, but possessed no detectable activity toward HVA. Both adrenal
gland and frontal cortex showed very low activity toward
all the acid metabolites.

Apparent K_m values for PST activity toward dopamine
and phenol were quite similar in the different tissues
(Rein et al.,1981c) (Table 6). Similar apparent K_m values and pH optima for HMPG in human platelet, gut and
kidney have also been reported by Anderson and Weinshil-

boum (1980). At present, therefore, there is no reason
to postulate that the two forms of PST differ from
tissue to tissue.

SELECTIVE INHIBITORS

The foregoing evidence pointing to the existence of
multiple forms of human PST raised the possibility of
distinguishing them by selective inhibition. Dichloronitrophenol (DCNP) and pentachlorophenol (PCP) have
been reported by Mulder and Scholtens (1977) to be
specific inhibitors in the rat of sulphate conjugation
as compared with glucuronide conjugation, both *in vitro*
and *in vivo*. These authors also showed that rat liver
PST activity toward harmol and phenol could be distin-
guished using DCNP, with harmol showing preferential
sensitivity to the inhibitor. PCP on the other hand
gave the same degree of inhibition with both substrates.
Drugs which modify sulphydryl groups have also been
used to distinguish between PST activity toward tyramine and p-nitrophenol in rat liver (Mattock and Jones,
1970).

In the light of such findings, it became of interest
to determine whether any of these drugs could selectively inhibit human platelet PST activity toward dopamine or phenol. The sulphydryl reagents, N-ethylmalei-
mide, p-chloromercuribenzoic acid, iodoacetic acid and
dithiothreitol were all ineffective at altering the
dopamine to phenol activity ratio. DCNP and PCP, however, caused a marked increase in this ratio (Rein et
al.,1981b). The dose-response curve for the inhibition
of platelet PST activity towards dopamine and phenol by
DCNP is shown in Fig.4a. It is clear that DCNP selectively inhibited the sulphoconjugation of phenol. Although similar results were obtained for PCP, it was somewhat less effective at distinguishing between the two
forms. DCNP was therefore used in all subsequent studies. Dose response curves with the platelet enzyme using tyramine, 5-HT and HMPG are shown in Fig.4b. These
substrates all gave similar inhibition curves to that
for dopamine, indicating that they are metabolized by
the same form of PST. The ability of DCNP to separate
the activities of the monoamines and their metabolites
from phenol is strong evidence for the existence of
multiple forms of PST and that the platelet contains
both forms of the enzyme.

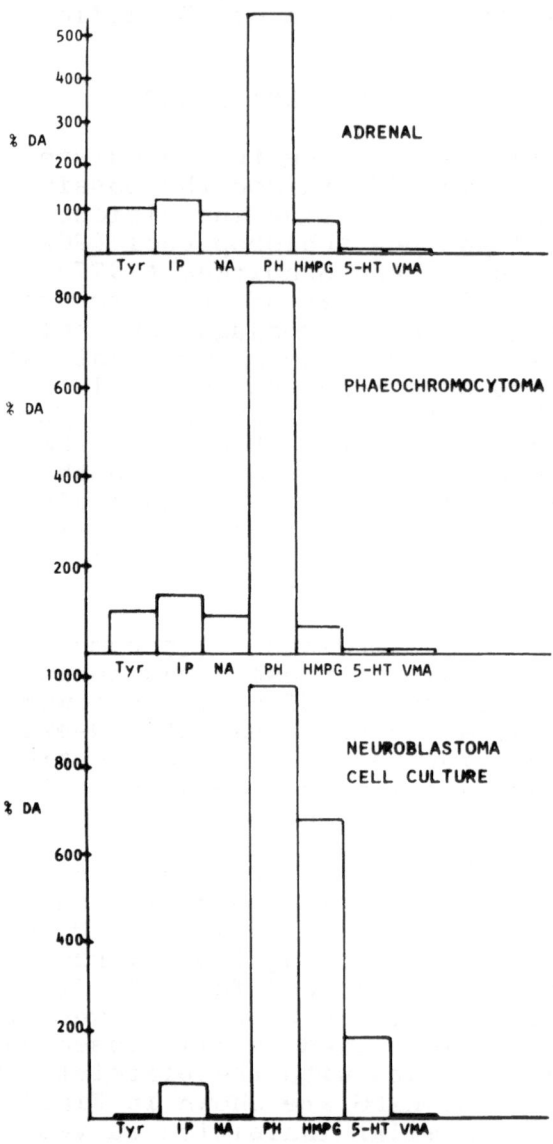

Figure 3

Substrate specificity pattern for PST in different human tissues.
Activity was assayed using a 30,000 g supernatant which was diluted 1:40 with 0.06% bovine serum albumin in the presence of 8 mM
dithiothreitol and 30 μM of each successive substrate. Adrenal gl-
and and phaeochromocytoma supernatants were obtained from two subjects and assayed independently. The results are mean values. Each
of the two confluent dishes of neuroblastoma cells was   assayed

separately and the results averaged. Results are expressed as in
Fig.2. For abbreviations, see caption to Fig.2 (from Rein et al.,
1981c and Rein and Sandler,1981).

TABLE 4. Substrate ratios in different human tissues

Tissue	n	Dopamine/ Tyramine ratio	Dopamine/ Phenol ratio	Dopamine/ HMPG ratio
Jejunum	4	1.3	5.6	5.1
Placenta	3	1.2	3.9	10.1
Platelet	4	1.1	2.5	4.9
Brain	8	2.5	1.0	1.2
Adrenal	3	1.0	0.12	1.3
Phaeochromo- cytoma	2	1.0	0.12	1.6
Neuroblas- toma cells	-	-	0.10	0.15

Ratio values were obtained from the data of Figures 2 and 3 expressed as pmoles product/min/mg protein. Brain and adrenal samples from different individuals were assayed separately and the
results averaged. n represents the total number of tissues from
different individuals assayed either as a pooled preparation or
separately.

TABLE 5. PST activity toward acid metabolites in different human
         tissues

Substrate	Platelet	Jejunum	Placenta	Adrenal	Brain
VMA	125	183	128	113	4.9
HVA	77	96	102	N.D.	N.D.
DOPAC	N.D.	22	N.D.	N.D.	N.D.

Relative activity (dopamine = 100)
PST activity was assayed on diluted human tissues, prepared as
described in Figure 2, using each substrate at 3 mM. Tissues from
3 subjects were pooled in all cases except for the adrenal which
was assayed in a pooled homogenate from 2 subjects. Results are
expressed relative to the specific activity obtained using 30 µM
dopamine. Each value represents the mean of 4 determinations.
N.D: not detectable (from Rein et al.,1981c).

TABLE 6. K_m values (µM) in different human tissues

Tissue	n	Dopamine	Phenol
Platelet	4	3.0	6.7
Jejunum	4	1.2	-
Placenta	3	2.8	-
Brain	3	3.1	10
Adrenal	2	-	5.5
Phaeochromocytoma	2	-	3.4

Apparent K_m values (µM) were determined using Lineweaver-Burk plots of PST activity obtained from pooled tissues prepared as described in Figure 2. n represents the total number of tissues from different subjects assayed as a pooled preparation. Similar values were obtained from 2-3 subsequent experiments (from Rein et al.,1981c).

The effect of DCNP on PST from human frontal cortex is shown in Figure 5. As with the platelet enzyme, this inhibitor could clearly distinguish between activity with dopamine and that with phenol (Rein et al.,1981b). In both tissues, the pattern with tyramine and 5-HT was also similar to that of dopamine. These results, therefore, conform with substrate ratio data from different tissues (Table 4) and indicate that monoamines are probably metabolized by the same form of PST. The situation with HMPG is somewhat less clear. Whereas in platelets the DCNP inhibition curve with HMPG was very similar to that with dopamine, in the frontal cortex it was somewhat shifted towards that of phenol. Further evidence that HMPG may not be metabolized solely by the dopamine-conjugating enzyme is that the dopamine/HMPG activity ratio varied considerably in different sites, with HMPG being a relatively better substrate in brain, adrenal and tumour tissues. It is therefore possible that HMPG is metabolized by a third form of the enzyme or, alternatively, that it may act as a mixed substrate with affinity for both the dopamine and phenol forms. Any interpretation of these data, however, must be made with caution for we do not know to what extent the properties of the various sulphotransferases have been altered during the process of postmortem deterioration. The platelet data, however, which were obtained from fresh tissue, clearly indicate that phenol and the monoamines are conjugated by distinct forms of PST.

INHIBITOR SENSITIVITY IN HUMAN PLATELETS

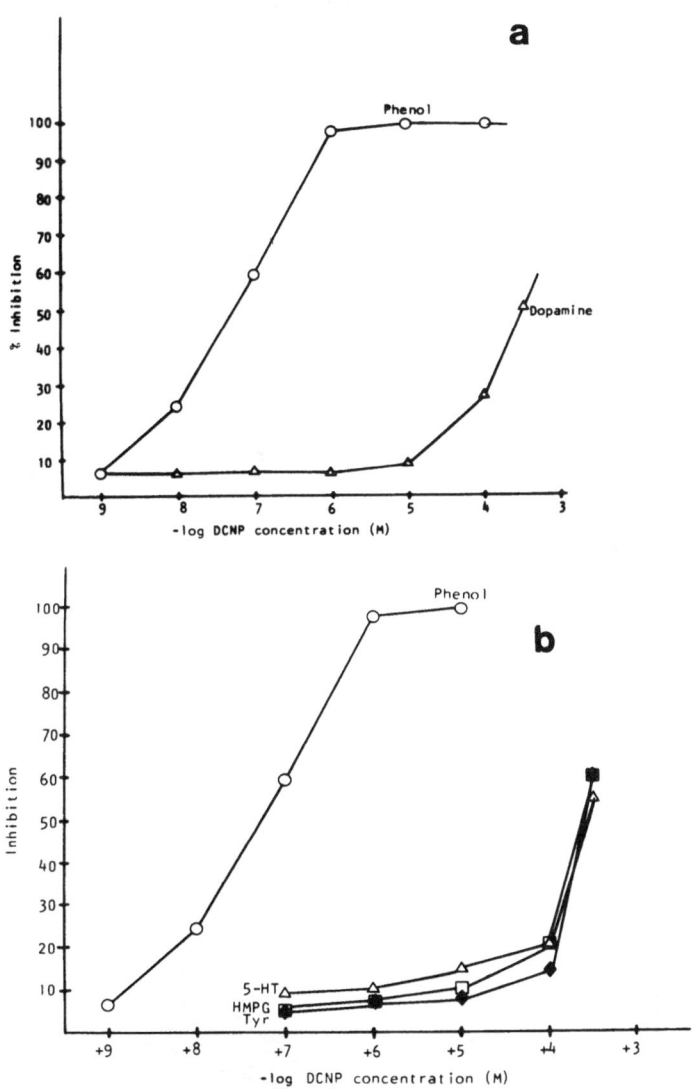

Figure 4
Sensitivity of platelet PST to inhibition by various concentrations of dichloronitrophenol (DCNP). PST activity was assayed in a pooled platelet homogenate from 4 subjects using 30 μM dopamine or phenol. DCNP was present during a 20 min preincubation and for the actual assay. Results are expressed relative to values obtained in the absence of DCNP. Each point represents the mean of 4 determinations (from Rein et al.,1981b).

114

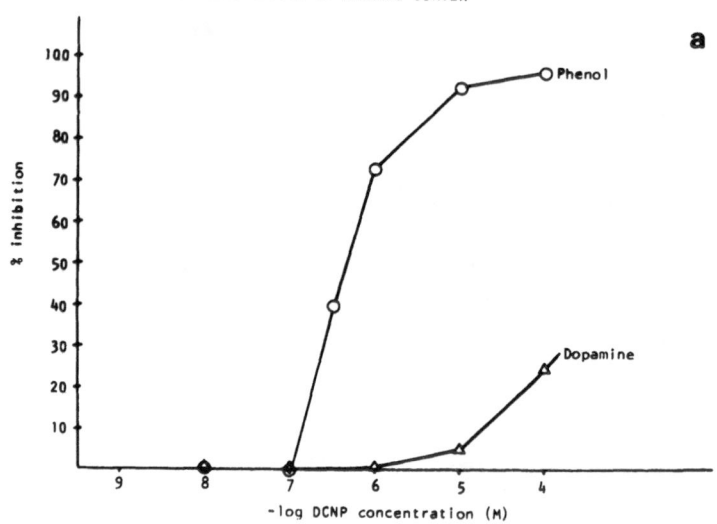

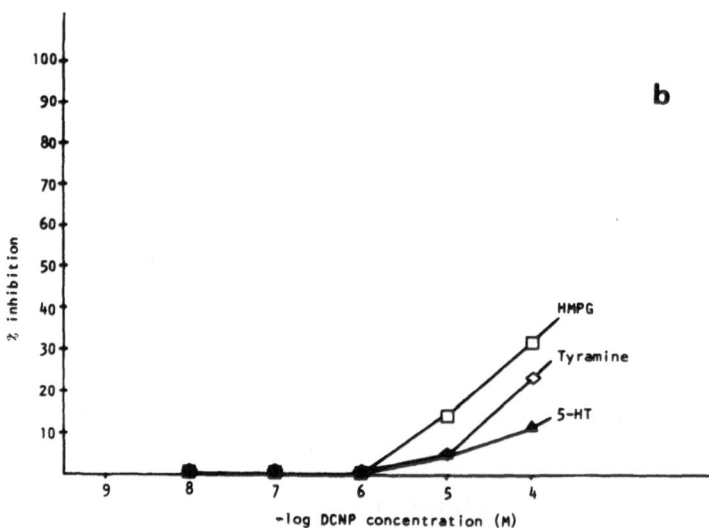

Figure 5
Sensitivity of frontal cortex PST to inhibition by various concentrations of DCNP. PST activity was assayed in a pooled
30,000 g supernatant from 3 subjects using 30 μM dopamine and
phenol in the absence of pargyline and dithiothreitol. Other conditions as for Figure 4 (from Rein et al.,1981b and d).

SPECIFIC ACTIVITY OF PST IN
DIFFERENT HUMAN TISSUES

Table 7 shows the specific activity of PST with DA and phenol in the different human tissues studied (Rein et al.,1981b). With both substrates, the jejunum was the most and the brain the least active tissue with the platelet giving intermediate values. The activity with the two substrates however was different in adrenal which showed substantially greater activity towards phenol. The high activity in the intestine confirms the view that sulphoconjugation is important in metabolizing exogenous as well as endogenous phenols, but the relatively low activity in the brain raises the question of whether the enzyme is functionally important in this tissue or not.

TABLE 7. PST specific activity toward dopamine and phenol in different human tissues

Tissue	n	Specific activity (pmoles/min/mg protein)	
		Dopamine	Phenol
Jejunum	4	138	24.5
Platelet	4	22.1	8.9
Adrenal	2	1.6	12.9
Placenta	3	1.6	0.41
Phaeochromo-cytoma	2	0.85	7.1
Brain	3	0.66	0.68

The relative activities in Figures 2 and 3 are expressed here as specific activities. All values either from 30,000 g supernatants or assayed in the presence of 0.06% bovine serum albumin (BSA) were corrected to the lower values that would have been obtained from diluted homogenates in the absence of BSA. This adjustment allows for direct comparisons between all tissues. n represents the total number of tissues from different subjects assayed either as a pooled preparation or separately. Each value is the mean of 4 determinations (from Rein et al.,1981b).

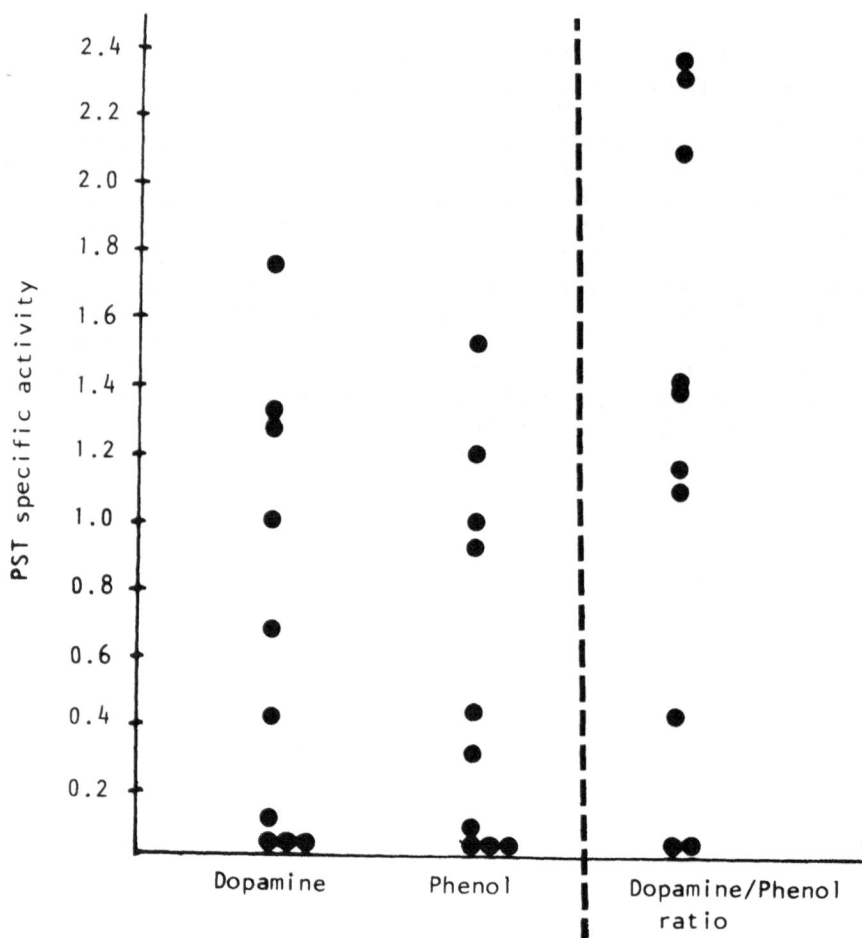

Figure 6
Individual variation in PST specific activity (pmoles/min/mg
protein)   towards 30 μM dopamine or phenol in frontal cortex
from 10 patients free from neurological disease at time of death.
A 30,000 g supernatant from each frontal cortex was assayed for
PST activity in the presence of 8 mM dithiothreitol. Each point
represents the mean of 4 determinations (from Rein et al.,1981d).

HUMAN BRAIN PST

Whereas all other tissues under scrutiny were either
procured fresh or by biopsy, brain tissue was obtained
postmortem. To ensure a minimum amount of deterioration, only brains stored less than 24 h prior to autopsy
were used. No correlation was observed between PST activity with either dopamine or phenol and postmortem de-
lay, a finding consistent with that of Foldes and Meek

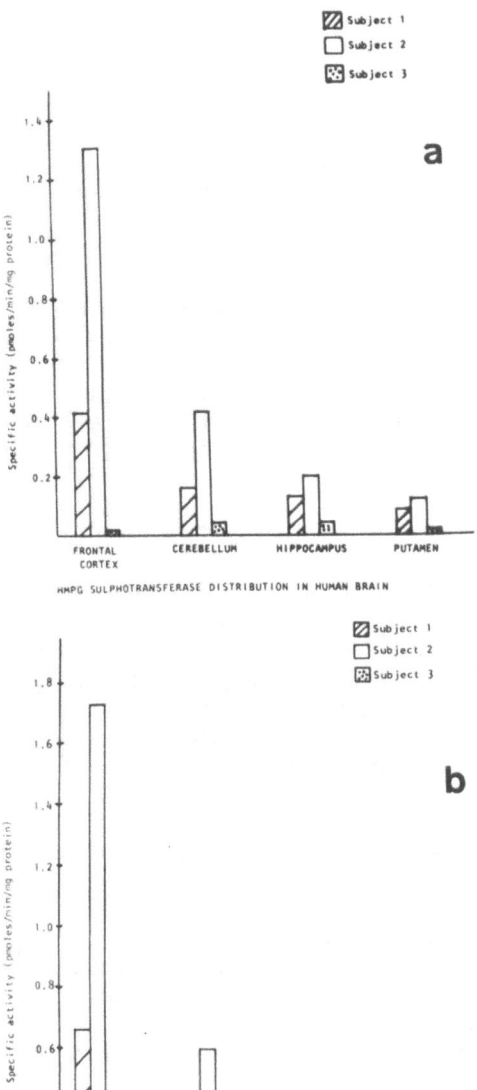

Figure 7

Distribution of PST activity toward 30 μM dopamine (a) and 1 mM
HMPG (b). A 30,000 g supernatant from each brain region was preincubated with 1 mM pargyline and assayed for PST activity in its
presence and that of 8 mM dithiothreitol. Each value represents
the mean of 4 determinations (from Rein et al.,1981d).

(1974). However, the possible contribution of post-mortem changes to low activities must be borne in mind. There was a large variation in PST activity toward both substrates in the 10 individual brains studied (Figure 6) and one-third of the brains analyzed had no detectable activity (Rein et al.,1981d). Whilst, on the surface of it, this finding points to much larger individual variation than has been noted with platelets, the physiological significance of these findings is unknown.

The distribution of PST activity toward dopamine and HMPG in different brain regions (Rein et al.,1981d) is presented in Figure 7. Frontal cortex showed the greatest activity followed by the cerebellum, hippocampus and putamen. A similar distribution was also obtained using 3 mM VMA, although only one brain had high enough activity for accurate analysis.

PST ACTIVITY IN NEURAL CREST TUMOURS

We have compared PST activity in phaeochromocytoma tumour tissue with normal adrenal gland. In two specimens obtained at operation, normal adjacent adrenal was provided together with tumour samples from the same patient. The results shown in Table 8 indicate that PST activity toward phenol in both tumour samples was decreased by a factor of approximately three compared with normal tissue.

TABLE 8. PST activity of phaeochromocytoma tumour tissue and normal adjacent adrenal

| | Dopamine | | Phenol | |
	Phaeochromo-cytoma	Adrenal	Phaeochromo-cytoma	Adrenal
Patient 1	1.5	0.80	2.4	8.9
Patient 2	0.42	4.2	13.6	30.9

Specific activity values for two patients whose phaeochromocytoma and normal adjacent adrenal were obtained at operation. PST was assayed by diluting a 30,000 g supernatant 1:40 in 0.06% BSA in the presence of 8 mM dithiothreitol and 30 µM of each substrate. Values which are the average of 4 determinations are not corrected for the presence of BSA as in Table 7 (from Rein and Sandler, 1981).

In one patient, PST activity toward dopamine was decreased ten-fold in the tumour samples, These preliminary results (Rein and Sandler,1981) suggest that tumour tissues are deficient in their ability to form sulphate conjugates, although the substrate specificity pattern (Figure 3) and the apparent K_m value for phenol (Table 6) were similar in both tissues. These results are consistent with the recent observation that human lung cancer in tissue culture preferentially forms glucuronide conjugates of phenolic substrates in contrast to normal lung cultures which predominantly give rise to sulphate conjugates (Mehta and Cohen, 1979). The clinical significance of these findings is at present unknown.

Although sulphate and glucuronide conjugates have been measured in rat and human cell lines (Michelot et al.,1977; Crooks et al.,1978), PST activity has not been previously reported in any tissue culture system. The use of actively dividing human neuronal cell lines which can be used as model systems for studying neuronal functions and which can be grown in adequate amounts for biochemical study has been gaining increasing popularity over the last decade. Several human neuroblastoma cell lines have been isolated which express certain biochemical properties associated with normal nerve cells (Schlesinger et al.,1976; West et al., 1977; Biedler et al.,1978). We are at present characterizing a new human neuroblastoma cell line which contains PST activity with similar properties to brain and adrenal enzyme (Rein and Sandler,1981). Further studies will be required to ascertain whether this cell line represents an adequate model for studying sulphate conjugation in normal human tissues.

DISCUSSION

This study provides strong evidence that phenol and catecholamines are metabolized by distinct enzymes in human tissues. The substrate specificity pattern in the different tissues suggests the possibility of a distinct PST preferentially active toward phenol. The activity ratio between dopamine and phenol further suggests that the various tissues examined have different relative amounts of the two enzymes with the adrenal and the brain containing relatively more of the the phenol-metabolizing enzyme. The selective inhibitor evidence confirms the presence of two distinct enzymes and shows that platelet and brain contain both forms. We propose calling the two forms PST M (for monoamine and monoamine metabolites) and PST P (for phenol). This evidence for the existence of two forms of PST is also supported by the data presented by

Sandler et al.(this volume) that platelet activity towards phenol and dopamine varies independently in different individuals. Both we and Weinshilboum et al. (this volume) have found that there is a high degree of correlation between the activities of dopamine, tyramine and HMPG in different individuals. Much more work is needed to delineate in greater detail the substrate specificity for the two forms. In platelets, for example, HMPG appears to be a substrate for PST M based on its sensitivity to DCNP and its covariation with other amines in different individuals. However, the dopamine to HMPG activity ratio both in adrenal biopsy samples and postmortem brain (Table 4) suggests that HMPG may be metabolized by a third form of PST or may, alternatively, be a substrate for PST P as well as for PST M.

These two forms of the enzyme in man may have some similarity to the two forms previously described in the rat (Mattock and Jones,1970; Barford and Jones, 1971; McEvoy and Carroll,1971). Two enzymes from rat liver, one specific for tyramine and tyrosine methyl ester and one for p-nitrophenol have been separated and shown to have differential sensitivity to drugs which modify sulphydryl groups. Jakoby and his colleagues(1980) have purified two groups of PST enzymes from rat liver with different pH optima, amino acid composition, antibody specificity and different relative activities toward phenols and monoamines. However, there are marked contrasts between rat and human PST, most notably in relative activities toward the monoamines and their metabolites. The rat liver and brain enzymes showed low activity towards dopamine, but substantial activity towards alcoholic and acidic metabolites (Foldes and Meek,1973; Meek and Neff,1973). In addition, as we show here, drugs affecting sulphydryl groups do not distinguish between the two human forms of the enzyme, whereas they do in the rat.

PST in all tissues has a much lower V_{max} than does MAO in the same site. However, the much lower K_m values for amines such as dopamine can mean that it is relatively more important than MAO when substrate concentrations are low. This point is illustrated in Figure 8 for   jejunal tissue and platelet. At concentrations of dopamine below 5 μM, one would expect conjugation to be more important than oxidation. This is supported by the evidence of Bronaugh et al. (1975 a, b) who have shown that the output of conjugated dopamine in man is relatively high compared with homovanillic acid for low oral doses of L-dopa whereas after

a large oral dose of L-dopa there is a higher proportion of homovanillic acid. Conjugation is the major
route of metabolism when small doses of ^{3}H-L-dopa are given
orally but O-methylation and oxidative deamination predominate when ^{3}H-L-dopa is given intravenously. These
data taken in conjunction with the *in vitro* enzyme findings, further support the conclusion that sulphate conjugation is important, at least in the periphery, for
metabolizing exogenous phenols and phenolic amines.

A similar kinetic analysis using data for the brain
indicates that oxidative deamination is considerably
more important than sulphate conjugation at all physiological substrate concentrations. However, it is possible that human brain PST is particularly unstable
after death and that its specific activity has been
considerably underestimated. Nevertheless, since several animal species, notably the squirrel monkey, also
appear to have extremely low PST activity in the brain
(Foldes and Meek,1974), it cannot be ruled out that
sulphate conjugation plays a relatively minor role under normal conditions. The distribution of PST in the
human brain does not follow that of the monoamines or
MAO, being higher in the frontal cortex than the putamen. Buu and Kuchel (this volume) have also reported
that rat brain areas with high concentrations of free
catecholamines do not have a correspondingly high PST
activity. These findings, therefore, raise the question
of the function of PST in the brain. Even so, its functional role may well have to be taken into account in
neuropsychiatric disorders where an 'amine imbalance
might exist in particular brain regions. It may also
assume greater importance following the administration
of monoamine oxidase inhibitors.

In order to assess the potential use of platelet PST
as a clinical marker for sulphate conjugation in the
brain and throughout the body, one needs to characterize and compare its biochemical properties in different human tissues. We report here a similar apparent K_m
for dopamine and phenol, and a similar substrate specificity pattern for the catecholamines in different
tissues. Previously reported findings indicated that
the gut, platelet and kidney enzymes have similar apparent K_m values and pH optima for HMPG (Anderson and
Weinshilboum,1980) and a significant correlation has
been observed between platelet and renal PST when assayed with HMPG in different individuals (Weinshilboum et
al.,this volume). The identification of M and P forms
of the enzyme may take us somewhat further along the

122 road and enable us to assess the activity of each form of the enzyme independently in different human disease states (see Sandler et al.,this volume).

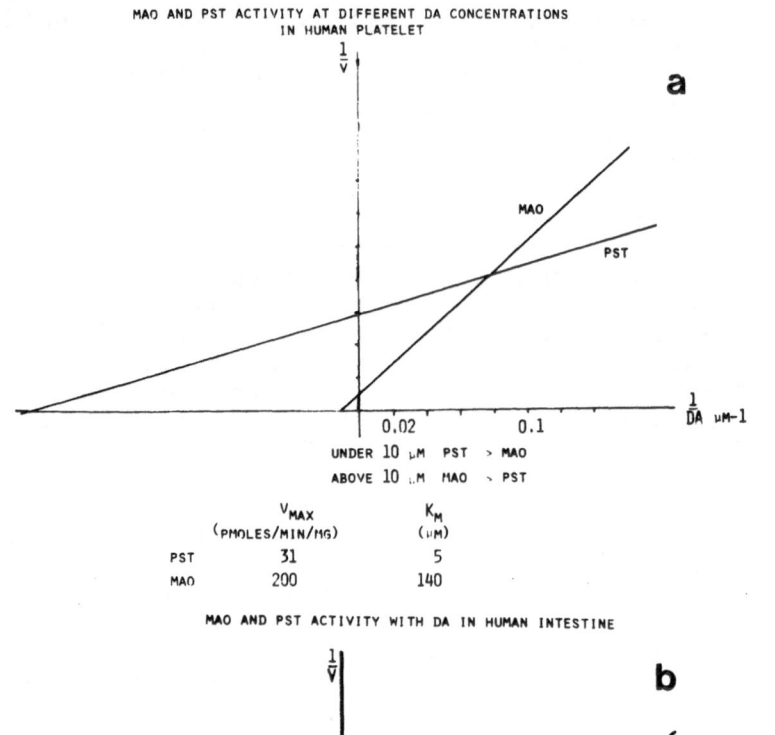

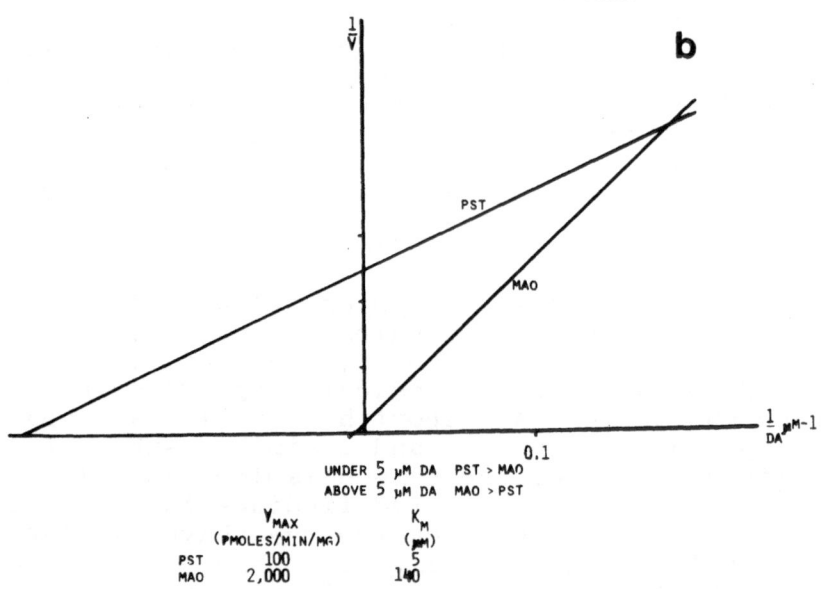

Figure 8
Kinetic analysis of MAO and PST in human (a) platelets and (b) intestine. PST data were obtained from Tables 3 and 6.

Adams,J.B.and Chulavatnatol,M.(1967). Enzymatic synthesis of steroid sulphates. IV. The nature of the two forms of estrogen sulphotransferase of bovine adrenals. Biochim.Biophys.Acta 146, 509-521.

Adams,J.B.and Poulos,A.(1976). Enzymatic synthesis of steroid sulphates. III. Isolation and properties of estrogen sulphotransferase from bovine adrenal glands. Biochim.Biophys.Acta 146, 493-508.

Anderson,R.J.and Weinshilboum,R.M.(1979). Phenolsulphotransferase: enzyme activity and endogenous inhibitors in the human erythrocyte. J.Lab.Clin.Med.,94, 158-171.

Anderson,R.J.and Weinshilboum,R.M.(1980). Phenolsulphotransferase in human tissue: radiochemical enzymatic assay and biochemical properties. Clin.Chim.Acta 103, 79-90.

Banerjee,R.K.and Roy,A.B.(1966). The sulphotransferases of guinea pig liver. Mol.Pharmacol., 2, 56-66.

Banerjee,R.K.and Roy,A.B.(1967). The formation of cholesteryl sulphate by androstenolone sulfotransferase. Biochim.Biophys.Acta 137, 211-213.

Barford,D.J.and Jones,J.G.(1971). Thiol-dependent changes in the properties of rat liver sulphotransferases. Biochem.J., 123, 427-434.

Baumann,E.(1876). Über gepaarte Schwefelsäuren in Organismus. Pflügers Arch.ges.Physiol.Menschen Thiere, 13, 285-308.

Biedler,J.L.,Roffler-Tarlov,S.,Schachner,M.and Freedman,L.S.(1978). Multiple neurotransmitter synthesis by human neuroblastoma cell lines and clones. Cancer Res.,38, 3751-3757.

Boström,H.and Wengle,B.(1967). Studies on ester sulphates: distribution of phenol and steroid sulphokinase in adult human tissues. Acta Endocr.,56, 681-704.

Bronaugh,R.L.,Hattox,S.E.,Hoehn,M.M.,Murphy,R.C.and Rutledge,C.O.(1975). The separation and identification of dopamine 3-O-sulfate and dopamine 4-O-sulfate in urine of parkinsonian patients. J.Pharmacol. Exp.Ther.,195,441-452.

Bronaugh,R.L.,McMurtry,R.J.,Hoehn,M.M.and Rutledge,C.O. (1975). Conjugation of L-dopa and its metabolites after oral and intravenous administration to parkinsonian patients.Biochem.Pharmacol.,24, 1317-1320.

124   Caldwell,J.(1976). The metabolism of amphetamines in mammals. Drug Metab.Rev., 4, 219-280.

Caldwell,J.(1980). Comparative aspects of detoxication in mammals. In Enzymatic Basis of Detoxication vol.1 (ed.W.B.Jakoby) Academic Press, New York, pp.85-114.

Crooks,P.A.,Breakefield,X.O.,Sulens,C.H.,Castiglione, C.M.and Coward,J.K.(1978). Extensive conjugation of dopamine (3,4-dihydroxyphenethylamine) metabolites in cultured human skin fibroblasts and rat hepatoma cells. Biochem.J.,176, 187-196.

Dodgson,K.S.(1977).Conjugation with sulphate. In Drug Metabolism - From Microbe to Man (eds.D.V.W.Parke and R.L.Smith), Taylor & Francis,London,pp.91-104.

Foldes,A.and Meek,J.L.(1973). Rat brain phenolsulphotransferase - partial purification and some properties. Biochim.Biophys.Acta 327, 365-373.

Foldes,A.and Meek,J.L.(1974). Occurrence and localization of brain phenolsulphotransferase. J.Neurochem., 23, 303-307.

Goodall,McC.and Alton,H.(1968). Metabolism of dopamine in human subjects. Biochem.Pharmacol.,17, 905-914.

Hart,R.F.,Renskers,K.J.,Nelson,E.B.and Roth,J.A.(1979). Localization and characterization of phenol sulphotransferase in human platelets. Life Sci.,24,125-130.

Hext,F.A.,Thomas,S.,Rose,F.C.and Dodgson,K.S.(1973). Determination and significance of L-tyrosine O-sulphate and its deaminated metabolites in normal human and mouse urine. Biochem.J.,134, 629-636.

Holcenberg,J.S.and Rosen,S.W.(1965). Enzymatic sulfation of steroids by bovine tissues. Arch.Biochem. Biophys. 110, 551-557.

Jakoby,W.B.,Sekura,R.D.,Lyon,E.S.,Marcus,C.J. and Wang, J.-L. (1980). Sulfotransferases. In Enzymatic Basis of Detoxication, vol.2 (ed.W.B.Jakoby) Academic Press, New York, pp.199-228.

Lewinsohn,R.,Glover,V.and Sandler,M.(1980). Development of benzylamine oxidase and monoamine oxidase A and B in man. Biochem.Pharmacol.,29, 1221-1230.

McEvoy,F.A.and Carroll,J.(1971). Purification from rat liver of an enzyme that catalyses the sulphurylation of phenols. Biochem.J.,123, 901-906.

Mattock,P.and Jones,J.G.(1970). Partial purification and properties of an enzyme from rat liver that catalyses the sulphation of L-tyrosyl derivatives. Biochem.J.,116, 797-803.

Meek,J.L.and Neff,N.H.(1973) Biogenic amines and their metabolites as substrates for phenol sulphotransferase (EC 2.8.2.1) of brain and liver. J.Neurochem., 21, 1-9.

Mehta,R.and Cohen,G.M.(1979). Major differences in the extent of conjugation with glucuronic acid and sulphate in human peripheral lung. Biochem.Pharmacol., 28, 2479-2484.

Michelot,R.J.,Lesko,N.,Stout,R.W.and Coward,J.K.(1977). Effect of S-adenosylhomocysteine and S-tubercidinylhomocysteine on catecholamine methylation in neuroblastoma cells. Mol.Pharmacol.,13, 368-373.

Mulder,G.J.and Scholtens,E.(1977). Phenolsulphotransferase and uridine diphosphate glucuronyltransferase from rat liver *in vivo* and *in vitro*. Biochem.J., 165, 553-559.

Nose,Y.and Lipmann,F.(1958). Separation of steroid sulfokinases.J.Biol.Chem.,233, 1348-1351.

Rein,G.,Glover,V.and Sandler,M.(1981a). Sulphate conjugation of biologically active monoamines and their metabolites by human platelet phenolsulphotransferase. Clin.Chim.Acta. In press.

Rein,G.,Glover,V.and Sandler,M.(1981b). Two forms of phenolsulphotransferase in human platelet and brain: selective inhibition by dichloronitrophenol.Submitted for publication.

Rein,G.,Glover,V.and Sandler,M.(1981c). Properties of phenolsulphotransferase M and P in different human tissues. In preparation.

Rein,G.,Glover,V.and Sandler,M.(1981d). Characterization of phenolsulphotransferase M and P from human brain. In preparation.

Rein,G.and Sandler,M.(1981). Sulphate conjugation of monoamines and their metabolites in neural crest tumours. In preparation.

Renskers,K.J.,Feor,K.D.and Roth,J.A.(1980). Sulfation of dopamine and other biogenic amines by human brain phenol sulfotransferase. J.Neurochem.,34,1362-1368.

126   Ruthven,C.R.J. and Sandler,M.(1969). The biosynthesis and metabolism of the catecholamines. In Progress in Medicinal Chemistry, vol.6 (eds.G.P.Ellis and G.B. West), Butterworth, London, pp. 200-265.

Schlesinger,H.R.,Gerson,J.M.,Moorhead,P.S.,Maguire,H. and Hummeler,K.(1976). Establishment and characterization of human neuroblastoma cell lines. Cancer Res., 36, 3094-3100.

Smith,E.R.B.and Weil-Malherbe,H.(1971). On the occurrence of glusulase hydrolyzable conjugates of 3,4-dihydroxyphenylacetic acid and homovanillic acid in human urine. Clin.Chim.Acta 35, 505-507.

Tipton,K.F.,Houslay,M.D. and Mantle,T.J.(1976). The nature and locations of the multiple forms of monoamine oxidase. In Monoamine Oxidase and its Inhibition (eds.G.E.W.Wolstenholme and J.Knight), Elsevier-Excerpta Medica-North Holland, Amsterdam, pp.5-16.

von Euler,U.S.,Floding,I. and Lishajko,F.(1959). The presence of free and conjugated 3,4-dihydroxyphenylacetic acid in urine and blood plasma. Acta Soc.Med. Upsalien 64, 217-225.

West,G.J.,Uki,J.,Herschman,H.R.and Seeger,R.C.(1977). Adrenergic, cholinergic, and inactive human neuroblastoma cell lines with the action-potential Na^+ ionophore. Cancer Res.,37, 1372-1376.

The Availability of Inorganic Sulfate for Sulfate Conjugation in Vivo

Gerard J Mulder and Klaas R Krijgsheld

Department of Pharmacology, State University of Groningen, Groningen,
The Netherlands

POOL SIZE OF INORGANIC SULFATE *IN VIVO*

For synthesis of the co-substrate of sulfation,
adenosine 3'-phosphate 5'-sulfatophosphate (PAPS), inorganic sulfate is required. Various factors affect
sulfate availability *in vivo* by changing the pool size
of inorganic sulfate. This will be reflected in changes
in the concentration of inorganic sulfate in the
blood.

The serum concentration of inorganic sulfate shows
wide species differences; in man it is approximately
0.3 mM, while in the rat and the mouse it is around
1.0 mM. Much higher concentrations, up to 2.5 mM are
found in the rabbit and the chicken (Mulder, 1981a;
Krijgsheld et al., 1980). Since the distribution volume of inorganic sulfate in various species, usually
measured with [^{35}S]-labelled sulfate ('radiosulfate
space'), is between 20 and 30% of body weight (Barrett
& Walser, 1969; Bauer et al., 1977; Mulder, 1981a),
the immediately available pool of inorganic sulfate
will vary more or less in parallel with the serum sulfate concentration. Table 1 illustrates the rather big
species differences in the pool size of sulfate. Little
is known about the inorganic sulfate content of tissues, since no assay method for sulfate in tissue homogenates has been developed. Only some indirect measurements are available; thus, the liver seems to contain approximately 0.4 µmol sulfate per g of liver that
can be used for sulfation (Mulder & Keulemans, 1978).

A second factor that determines the pool size is
the rate at which inorganic sulfate is generated; this
newly synthesized sulfate either makes up for a loss of

127

TABLE 1

Pool size of inorganic sulfate in various species

Species	Serum sulfate (mM)	Sulfate distribution volume (% of body wt.)	Directly available sulfate (μmol/kg)	Urinary excretion of sulfate (μmol/kg/hour)
Rat	0.9	30	270	70
Rabbit	2.0	25	500	25
Mouse	1.2	25	300	-
Man	0.3	20	60	15
Dog	1.4	20	280	25

The original references used for the various values can be found in Mulder, 1981b.

sulfate as a result of sulfation reactions, or is excreted in urine when the supply is in excess. Under normal conditions inorganic sulfate is excreted in urine, although the kidneys reabsorb inorganic sulfate (Mudge et al., 1973). Thus, when inorganic sulfate in the body is partly depleted by a high dose of a substrate for sulfation, the urinary excretion of sulfate is strongly decreased, while still sulfate is present in the serum (Büch et al., 1968; Krijgsheld et al., 1981). Indeed, a moderate dose of such a substrate will not decrease serum sulfate, presumably because a decreased urinary excretion compensates for the loss of sulfate by conjugation. The origin of this urinary excreted sulfate is the catabolism of sulfur-containing substances in the body. This includes both sulfate produced from cysteine taken up with the food, and sulfate that arises from the break-down of e.g. endogenous proteins (via cysteine) and glycosaminoglycans. The urinary excretion of sulfate, therefore, reflects a complex situation in which catabolism, food and composition of the food each play a role. An increased supply of cysteine and methionine will lead to increased urinary sulfate levels (Mulder, 1981a; Sabry et al.

1965; Bodwell et al., 1978; Stipanuk, 1979). But also
an imbalanced amino acid composition of the food may
increase urinary sulfate: when threonine becomes rate
limiting, the relative excess of cysteine and methionine is sulfoxidized, so that urinary sulfate increases (Lakshmanan et al., 1978). When rats are fasted,
urinary sulfate excretion decreases rapidly, because
the catabolism of endogenous compounds will gradually
yield less sulfate (Krijgsheld et al., 1981); since
no new cysteine is supplied the sulfate availability
decreases, even though serum sulfate in the rat is maintained for at least 3 days after start of the fast
(Table 2).

In conclusion, the pool size of inorganic sulfate
is very species dependent, and can easily be manipu-
lated by dietary regimens. Furthermore, the serum sulfate concentration seems very sensitive to stress factors, probably because glomerular filtration rate may
be one of the main regulatory factors (Krijgsheld et
al., 1981; and unpublished data).

SOURCES OF INORGANIC SULFATE IN THE FOOD

Probably the main natural source of inorganic sulfate *in vivo* is cysteine. It is either absorbed as
such, or it is generated from methionine by transsulfurylation. The SH-group of cysteine is oxidated,
which yields sulfite through various intermediates
(Singer, 1975). This is converted by sulfite oxidase
to inorganic sulfate. A genetic deficiency in sulfite
oxidation in man has been detected. This defect leads
to neurological disorders and, ultimately, to death
(Irreverre et al., 1967; Rosenblum, 1968; Shih et
al., 1977). The dietary treatment, a diet low in cysteine and methionine, alleviates the symptoms; however, probably it is advisable to add inorganic sulfate to provide sufficient sulfate for the necessary
sulfation reactions.

It has been almost completely overlooked in Pharmacology that inorganic sulfate salts are rapidly and
completely absorbed when they are given orally, unless
doses are administered that are too high and cause
diarrhea (Dziewiatkowski, 1949; Bauer, 1976; Krijgsheld
et al, 1979; Mulder, 1981a). Therefore, the explanation
for the laxative action of sulfate, namely that it is
not absorbed from the gut and thus 'pulls' water into
the lumen of the gut by osmotic action, clearly is un-
tenable in view of the large body of evidence in the
literature that shows excellent absorption of sulfate
(Krijgsheld et al., 1979). At present it is not clear

TABLE 2

Effect of fasting on serum sulfate concentration and the urinary excretion of inorganic sulfate in the rat.

Days fasted	Serum sulfate (mM)		Urinary sulfate (μmol/24 hr)	
	Fasted	Control	Fasted	Control
1	1.00 ± 0.06 (4)*	0.85 ± 0.02 (4)	165 ± 12 (6)*	334 ± 21 (6)
2	0.76 ± 0.01 (8)	0.83 (2)	168 ± 19 (5)*	384 ± 31 (6)
3	0.94 ± 0.00 (4)	0.83 ± 0.06 (4)	224 ± 33 (3)*	319 ± 58 (3)

Data are taken from Krijgsheld et al. (1981). The rats (200 g bodywt initially) were caged individually in metabolism cages. Blood was taken between 10 and 12 a.m.; each rat was used only once for taking a blood sample and sulfate was determined in serum and in the urine collected during the 24 preceeding hrs. Controls were fed ad libitum. The means + SEM are given, and the number of rats used is given in parentheses. *Means signi- ficantly different from control at p < 0.05 (Wilcoxon).

how much of the body requirement of sulfate is provided with the food in the form of absorbed inorganic sulfate. There is no doubt, however, that the addition of inorganic sulfate to the food causes increases in serum sulfate, and may have a sparing effect on the requirement of cysteine and methionine in several species, including man (Baker, 1976; Zezulka & Calloway, 1976; James & Hove, 1978; Mulder, 1981a).

Non-physiological sulfur-containing compounds can be used to increase the inorganic sulfate concentration in vivo. Thus, for instance, D-cysteine leads to a rapid increase of serum sulfate after oral administration (Glazenburg et al., 1981). The effects of oral doses of D-cysteine and L-cysteine on serum sulfate in the rat are shown in Figure 1. As yet, similar data are not available in man.

UPTAKE OF SULFATE BY THE TISSUES

Little is known of the uptake of inorganic sulfate by the various tissues. The available reports have utilized [35S]-labelled sulfate, and show that most, probably all, tissues incorporate [35S]-sulfate after its intravenous administration (Dziewiatkowski, 1958; Lindahl & Höök, 1978). However, it is not known whether the uptake processes are rapid and lead to equilibration or to accumulation of sulfate in the cells, as compared to the serum concentration of sulfate. A fractional distribution of radio-labelled sulfate over various tissues shows that the uptake may be very different but that more than 80 per cent is located extracellular (Barrett & Walser, 1969). The fact that the volume of distribution of radiosulfate shows a slow time-dependent increase probably reflects the slow exchange and uptake of radiosulfate in several tissues, as compared with a rapid equilibration with some other tissues (Barrett & Walser, 1969; Bauer et al., 1977; Thornton & English, 1977).

For the liver the data indicate a rapid equilibration of sulfate in serum with that in the liver (Herbai, 1970; Mulder & Scholtens, 1978), probably within a couple of minutes (Mulder & Scholtens, 1978). Studies with isolated hepatocytes from the rat showed intracellular accumulation of sulfate at low sulfate concentration in the incubation medium, while at physiological sulfate levels a rapid equilibration between medium and cells occurred (Schwarz, 1980a), and no accumulation over the concentration in the medium.

The uptake of sulfate in the brain seems a slow process, which requires several hours for equilibration

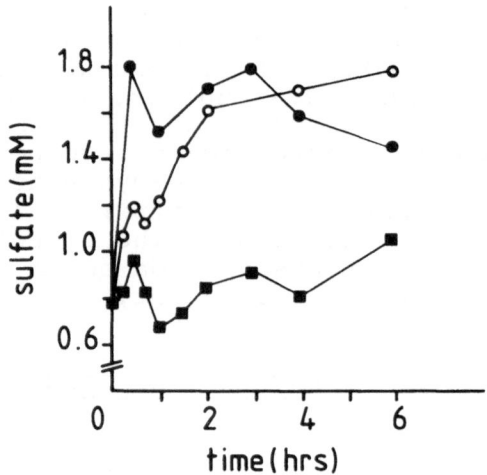

Figure 1. Effect of L-cysteine
and D-cysteine on serum inorganic sulfate in the rat

Data are from Glazenburg et al. (1981). Rats
received an oral dose of 8 mmol/kg L-cysteine
(o - o) or D-cysteine (● - ●). The controls
(■ -■) received an oral dose of 7.7 mmol/kg
sodium chloride, which was also present in
the cysteine doses. Various times after administration blood was taken by cardiac puncture from four rats at each point in time,
and serum sulfate was determined; the rats
were used only once. The means are given in
the Figure. The SEM was in most cases below
5 % of the mean, but always below 10 % of
the mean value.

(Dziewiatkowski, 1970; Drewes & Gilboe, 1977). Brain
synaptosomes from sheep take up inorganic sulfate by a
saturable transport with a K_m of 4.4 mM (White, 1979).

Finally, sulfate uptake in erythrocytes and ascites tumor cells has been extensively studied, and shown
to be a sodium-dependent carrier transport, that is
also involved in Cl^-/SO_4^{2-} exchange and SO_4^{2-}/SO_4^{2-} self-
exchange (Cabantchik et al., 1978; Levinson, 1978).
These data have more extensively been reviewed elsewhere (Mulder, 1981a).

ACTIVATION OF SULFATE

Once inorganic sulfate is available in the cells,
it is converted to adenosine 3'-phosphate 5'-sulfato-
phosphate (PAPS) in a two step reaction, catalyzed by
ATP-sulfurylase and APS-kinase, respectively. Sulfate
activation has recently been reviewed (Mulder, 1981b).
Although much information has accumulated about the
first step of this reaction, the step catalyzed by APS
kinase has been very little studied.

An important characteristic of the reactions is
that the thermodynamics of the first step are very un-
favourable towards formation of APS. Therefore, the
second reaction is required to make the formation

$$SO_4^{2-} \quad + \quad ATP \rightleftharpoons APS \quad + \quad PP_i$$

$$APS \quad + \quad ATP \rightleftharpoons PAPS \quad + \quad ADP$$

of PAPS possible. Yet, this situation probably
ensures that the steady state concentration of PAPS
is low, which seems an advantage, because this compound
is rather metabolically labile (Mulder, 1981b). Only
one publication is available so far, that gives information about cellular concentrations of PAPS. In rat
liver the concentration is 29 μM, but in brain no PAPS
could be demonstrated (Wong & Yeo, 1979).

Yet, the rate at which PAPS can be synthesized, at
least in the liver, is very high, as long as sulfate
is available in the blood or perfusion medium: a rate
of 0.1 $\mu mol/min/g$ of liver could be maintained for 60
minutes in the single-pass perfused rat liver (Pang et
al., 1981). Indeed, it seems that PAPS in the liver is
in rapid equilibrium with inorganic ($[^{35}S]$-labelled)
sulfate in blood (Mulder & Scholtens, 1978). However,
there is no information on the regulation of PAPS concentration in the liver and other tissues.

Obviously, intravenous or intraperitoneal injection of inorganic sulfate leads to an increased availability of sulfate in blood and, therefore, in the tissues, the latter depending on the rate of uptake. This
increase most likely is only of relatively short duration, because of rapid elimination of sulfate by the
kidneys (Mudge et al., 1973). A more prolonged increase
of the serum sulfate level will be attained when a sulfate salt is given orally. The serum sulfate concentration increases from 0.9 mM normally to as much as
2.0 mM for several hours after a high oral dose of sulfate in the rat (Krijgsheld et al., 1979). When sulfate
is added to the food (0.8 % w/w of sodium sulfate) the
serum sulfate concentration in rats showed a moderate increase from 0.96 ± 0.10 mM in controls to 1.29 ± 0.15 mM
in the treated group (n = 6; means ± SEM). A too high sulfate supply may affect the calcium availability *in vivo*,
because it stimulated the urinary excretion of calcium
(Whiting & Draper, 1980).

The administration of L- and D-cysteine results in
enhanced sulfate levels for a prolonged time when these
amino acids are administered orally. The advantage of
D-cysteine is that it will not stimulate protein synthesis (at least not directly), since conversion of D-
cysteine to L-cysteine presumably occurs only very
slowly (via D-amino acid oxidase and a transaminase).
Furthermore, D-cysteine is not converted to taurine
(Glazenburg et al., 1981), so that also the taurine
conjugation is not affected. With L-cysteine administration not only inorganic sulfate will be increased,
but also taurine, glutathione and the rate of protein
synthesis (Stipanuk, 1979; Glazenburg et al., 1981;
Tateishi et al., 1972).

In man only limited data are available on serum
sulfate and administration of sulfate or cysteine. Presumably these will increase serum sulfate and thereby
sulfation (Mulder, 1981a; Büch et al., 1968; Levy &
Matsuzawa, 1967; Houston & Levy, 1976; Bonham Carter
et al., 1980), but only indirect findings are available.

Thus, for short-term increases of serum sulfate
oral administration of inorganic sulfate will be sufficient for a few hours. But when sulfate supply should
be increased for a more prolonged period, for instance
days, a repeated oral administration of L- or D-cysteine may be the method of choice, with some preference
for D-cysteine. The toxicity of long-term administration of D-cysteine has not yet been evaluated.

Already in 1952 Bray et al. (1952) noted that in-organic sulfate could be depleted by high doses of sub-strates for sulfation. They administered various sul-fate precursors orally to rabbits, and demonstrated that this addition increased the sulfation of the phe-nols used in their studies; the product of the compe-ting reaction, glucuronidation, was decreased under those conditions. However, they did not prove unequivo-cally that depletion of sulfate was the reason for de-creased sulfation of the phenols because they did not measure the sulfate concentration in the serum of the rabbit, although, in retrospect, the sulfate require-ment in their experiments was high enough to account for total sulfate depletion, assuming a serum sulfate concentration of 2.0 mM (Krijgsheld et al., 1980).

Later several reports confirmed that additional sulfate increased sulfation of various compounds *in vivo* (Levy & Matsuzawa, 1967; Houston & Levy, 1976; Ga-linsky et al., 1979; Mulder & Meerman, 1978; Bonham Carter et al., 1980; Mulder, 1981a). Although the doses of the substrates employed in these studies were much lower than those in the study of Bray et al. (1952), a similar explanation was suggested for the fact that co-administration of sulfate-precursors enhanced sulfation. Yet, data on serum sulfate were still lacking.

Indeed, drugs like paracetamol (Krijgsheld et al., 1981), salicylamide (Greiling & Schulder, 1963) and phenol (Weitering et al., 1979) do decrease serum sul-fate and these data confirm that relatively mild doses of drugs that are sulfated may partly deplete sulfate in serum (and the tissues !?). As a tool to deplete serum sulfate, however, they are usually rather unat-tractive, since the unconjugated drug is still present in the blood, while serum sulfate already starts to recover (Figure 2). Therefore, any effect of such a pretreatment may be due to both sulfate depletion and the presence of a competitor for sulfation (e.g. para-cetamol in Figure 2).

An alternative method is the use of a diet low in sulfur-containing amino acids, such as a casein diet (Table 3). Obviously under those conditions little cys-teine will be 'left over' for sulf-oxidation, so that not only urinary, but also serum sulfate decreases pro-foundly (Krijgsheld et al., 1981). Yet the animals still may get enough cysteine and methionine to main-tain growth, so that no totally catabolic state re-sults as occurs during fasting (Table 2). Although serum sulfate decreases within a few days on the low-

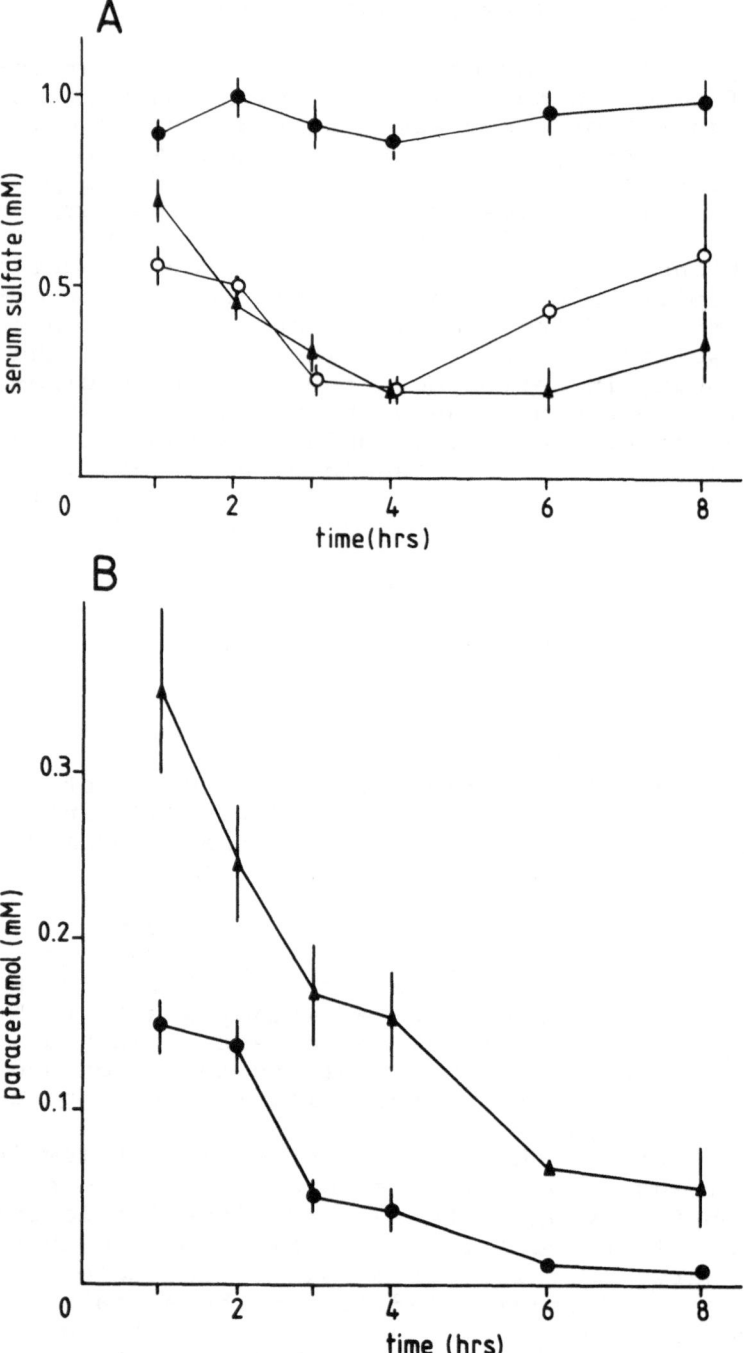

sulfur diet, yet adaptive changes in the conjugating enzymes may occur, that complicate the interpretation of findings on, for instance, the balance between glucuronidation and sulfation of a substrate *in vivo* (Magdalou et al., 1979).

A third method to decrease serum sulfate was found quite unexpectedly when oral administration of saline (0.9 % sodium chloride in water) caused a significant decrease of serum sulfate (Krijgsheld et al., 1980, 1981). However, the latter is no more than 30 % decrease at a high dose of sodium chloride, so that the effect is rather limited. The mechanism most likely is an inhibition of reabsorption of sulfate in the kidneys (Berglund & Lotspeich, 1956).

There is no method to deplete serum sulfate without some draw-back; it depends on the experiment, whether one or the other method will be preferred. Clearly, the analysis will be complicated, and more than one effect may contribute to the overall-result.

THE DEPENDENCE OF SULFATION RATE ON SERUM SULFATE CONCENTRATION

It seems reasonable to assume that there is a direct correlation between serum sulfate and sulfate concentration in the various tissues, although the linking factor may be very different for different tissues. It can not be excluded, however, that a selective depletion of sulfate may occur in one or more tissues, while the serum concentration is still hardly affected; the reverse situation, of course, is also possible. As yet no data are available on these aspects. Since sulfation takes only place inside cells, sulfate must become available intracellularly, both for activation to PAPS, and for the transfer reaction.

Figure 2. The effect of oral administration of acetaminophen on serum sulfate in the rat, and the plasma disappearance of acetaminophen. Data are taken from Krijgsheld et al. (1981). An oral dose of 1 (o, o) or 1.5 (▲ - ▲) mmol of acetaminophen per kg bodywt was administered at t = 0 by gastric tube. At different time intervals blood was collected from the aorta under ether anesthesia. Inorganic sulfate and unconjugated acetaminophen were determined in the sera; each point represents the mean (± SEM) of 4 rats. Panel A shows the data for inorganic sulfate in serum, panel B those for acetaminophen (● at 1.0 mmol/kg and ▲ at 1.5 mmol/kg).

TABLE 3

Effect of a low-protein diet on serum sulfate and
 urinary inorganic sulfate excretion in the rat

Days on diet	Serum sulfate (mM)		Urinary inorganic sulfate (μmol/kg/24hr)	
	Diet	Control	Diet	Control
1	0.83 ± 0.06	0.96 ± 0.02	910 ± 55	1880 ± 40
2	0.53 ± 0.06	0.96 ± 0.04	345 ± 25	1800 ± 85
3	0.54 ± 0.06	0.93 ± 0.10	215 ± 30	1990 ± 290
4	0.47 ± 0.07	0.99 ± 0.03	130 ± 20	1750 ± 85

Data are taken from Krijgsheld et al. (1981).
The rats (200 g bodywt initially) were caged
individually in metabolism cages, and provided
with either normal rat food, or an 8 % casein
diet that contained the normal composition in
terms of vitamins and trace elements. The sulfur content of the casein diet was about one-
third that of the control diet; the rats consumed the same amount of each diet, so that
the control rats consumed about three times
more sulfur than the diet rats. Between 10 and
12 a.m. a blood sample was taken; analysis
was as described in the legend to Table 2.

The fact that an increased sulfate supply does increase sulfation *in vivo* of several drugs under certain
conditions (Mulder, 1981a; Büch et al., 1968; Levy &
Matsuzawa, 1967; Houston & Levy, 1976; Bonham Carter et
al., 1980; Bray et al., 1952; Galinsky et al., 1979;
Mulder & Meerman, 1978) indicates that this intracellular sulfation is sensitive to increase in the serum
sulfate concentration.
Little is known about the sulfate-dependence of
sulfation rate. In the perfused rat liver the sulfation
of harmol seemed to require rather high levels of inorganic sulfate in the perfusion medium before saturation
of sulfation occurred: the physiological sulfate concentration of 1.0 mM probably is not yet saturating (Mulder & Keulemans, 1978). No further experiments of this
type have yet been reported, nor *in vivo* studies in

which the serum sulfate concentration has been varied in a controlled fashion. Two explanations are possible for a decrease of sulfation at increasing dose of a substrate *in vivo*. On one hand, the overall-sulfation process (both activation of sulfate and sulfate transfer) may have a rather high K_m for inorganic sulfate, of the order of the serum concentration, in which case any increase or decrease of serum sulfate will, respectively, increase or decrease the sulfation rate because of 'enzyme-kinetic' reasons. On the other hand, the K_m for sulfate may be very low, for instance of the order of 1 % of the serum concentration, in which case the rate of sulfation will not be affected by a decrease of serum sulfate, or an increase, unless serum sulfate becomes completely depleted. As yet there are no indications that one or the other system operates *in vivo*, because the type of experiment required, i.e. steady state infusions of substrate and sulfate, have not been performed. Both mechanisms may explain most of the data published so far.

Isolated hepatocytes incubated at different sulfate concentrations have also been used. Unfortunately, very different apparent K_m values for sulfate were found. Dependent on the substrate, saturation of sulfation required a sulfate concentration of 10 mM for paracetamol (Moldeus et al., 1979), 200 µM for 1-naphthol (Schwarz, 1980a) and at least 5 mM for harmol (Koster et al., 1981). Furthermore, the isolated hepatocytes have to be preincubated at $37^{o}C$, to deplete them of all their endogenous sulfate and/or PAPS (Koster et al., 1981). This wide variety may have several reasons. As yet there are no data suggesting the existence of more than one intracellular pool of PAPS, or the presence of multiple forms of the sulfate activating enzymes, ATP-sulfurylase and APS-kinase. Therefore, one would expect that the K_m for sulfate of the PAPS-synthesizing machinery (Figure 3) would be the same for every substrate of sulfation, if the latter does not affect the activating enzymes. Since there is clearly multiciplicity of sulfating enzymes, even of the phenol sulfotransferases (Roy, 1981; Jakoby et al., 1980) substrates sulfated by different transferases might show a different K_m for sulfate, because the sulfotransferases involved may have a different K_m for PAPS. On the other hand, the PAPS supply at low requirement for sulfation (i.e., at low sulfation rate) may be saturating at low sulfate levels, whereas at high sulfation rate a higher rate of PAPS synthesis, and therefore, a higher sulfate concentration may be required. The latter would imply that the apparent K_m

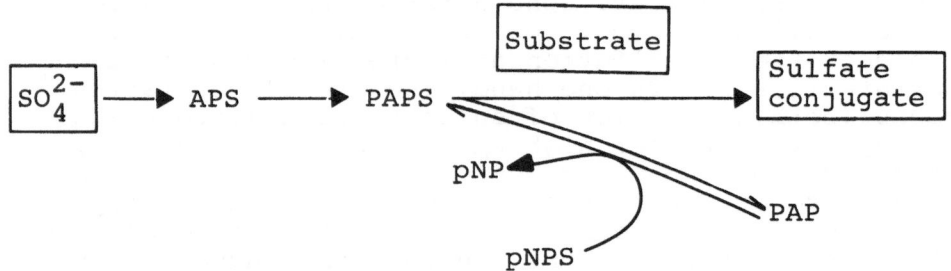

Figure 3. Components involved in the overall-sulfation pathway APS is adenosine 3'-phos-phate; PAP is adenosine 3'-phosphate 5'-phos-phate; pNP is p-nitrophenol; pNPS is p-nitro-phenyl sulfate.

for sulfate of the overall-sulfation process may depend on the rate of sulfation required at a certain substrate concentration of the acceptor. As yet, no definite con-clusion can be drawn, and more data are needed. The finding that p-nitrophenyl sulfate strongly stimulated the sulfation of 1-naphthol in rat hepatocyte incubat-ions suggests that indeed the activation of sulfate to PAPS may be limiting in the sulfation process (Schwarz, 1980b).

REFERENCES

Baker, D.H. (1976). Nutritional and metabolic interre-lationships among sulfur compounds in avian nutri-tion. Fed. Proceed. 35, 1917-1922.

Barrett, T.M. and Walser, M. (1969). Extracellular fluid in individual tissues and in whole animals: distribu-tion of radiosulfate and radiobromide. J. Clin. Invest 48, 56-66.

Bauer, J.H. (1976). Oral administration of radioactive sulfate to measure extracellular fluid space in man. J. Appl. Physiol. 40, 648-650.

Bauer, J.H., Burt, R.W., Whang, R. and Grim, C.E. (1975). Simultaneous determination of extracellular fluid and total body water. J. Lab. Clin. Med. 86, 1003-1017.

Berglund, F. and Lotspeich, W.D. (1956). Renal tubular reabsorption of inorganic sulfate in the dog, as af-fected by glomerular filtration rate and sodium chlo-ride. Amer. J. Physiol. 185, 533-538.

Bodwell, C.E., Schuster, E.M., Brooks, B. and Womack, M. (1978). Biochemical indices in humans of protein nu-tritive value. Nutr. Rep. Internatl. 18, 125-133.

Bonham Carter, S.M., Goodwin, B.L., Sandler, M., Gillman, P.K. and Bridges, P.K. (1980). Decreased conjugated tyramine output in depression: the effect of oral L-cysteine. Brit. J. Clin. Pharmacol. 10, 305-308.

Bray, H.G., Humphris, B.G., Thorpe, W.V., White, K. and Wood, P.B. (1952). Kinetic studies of the metabolism of foreign organic compounds. 4. The conjugation of phenols with sulphuric acid. Biochem. J. 52, 419-423.

Büch, H., Rummel, W., Pfleger, K., Eschrich, C. and Texter, N. (1968). Ausscheidung freien und konjugierten Sulfates bei Ratte und Menschen nach Verabreichung von N-Acetyl-p-Aminophenol. Naun. Schm. Arch. Pharmakol. 259, 276-289.

Cabantchik, Z.I., Knauf, P.A. and Rothstein, A. (1978). The anion transport system of the red blood cells. Biochim. Biophys. Acta 515, 239-302.

Drewes, L.R. and Gilboe, D.D. (1977). Nutrient transport systems in dog brain. Fed. Proceed. 36, 166-170.

Dziewiatkowski, D.D. (1949). On the utilization of exogenous sulfate sulfur by the rat in the formation of ethereal sulfates as indicated by the use of sodium sulfate labelled with radioactive sulfur. J. Biol. Chem. 178, 389-393.

Dziewiatkowski, D.D. (1958). Autoradiographic studies with ^{35}S-sulfate. Internatl. Rev. Cytol. 7, 159-189.

Dziewiatkowski, D.D. (1970). Metabolism of sulfate esters. In Symposium Sulfur in Nutrition (eds O.H. Muth and J.E. Oldfield), p. 97-125. Avi Publ. Comp., Westport, Conn.

Galinsky, R.E., Slattery, J.T. and Levy, G. (1979). Effect of sodium sulfate on acetaminophen elimination by rats. J. Pharmaceut. Sci. 68, 803-805.

Glazenburg, E., Krijgsheld, K.R., Scholtens, E. and Mulder, G.J. (1981). The effect of oral D-cysteine and L-cysteine on serum sulfate concentration and sulfate availability for conjugation in the rat in vivo. Submitted.

Greiling, H. and Schuler, B. (1963). Zur Wirkungsweise der Salizylsäure, Azetylsalizylsäure und des Salizylamids. Z. Rheumaforsch. 22, 47-56.

Herbai, G. (1970). A double isotope method for determination of the miscible inorganic sulfate pool of the mouse applied to in vivo studies of sulfate incorporation into costal cartilage. Acta Physiol. Scand. 80, 470-491.

Houston, J.B. and Levy, G. (1976). Drug biotransformation interactions in man VI: acetaminophen and ascorbic acid. J. Pharmaceut. Sci. 65, 1218-1221.

142   Irreverre, F., Mudd, S.H., Heizer, W.D. and Laster, L. (1967). Sulfite oxidase deficiency: studies of a patient with mental retardation, dislocated ocular lenses, and abnormal urinary excretion of S-sulfo-L-cysteine, sulfite and thiosulfate. Biochem. Med. 1, 187-217.

Jakoby, W.B., Sekura, R.D., Lyon, E.S., Marcus, C.J. and Wang, J.L. (1980). Sulfotransferases. In The Metabolic Basis of Detoxication (eds. W.B. Jakoby, J. Bend and J. Caldwell), vol. 2, chapter 11. Academic Press, New York, N.Y.

James, K.A.C. and Hove, E.L. (1978). Effects of inorganic sulfate supply on the growth of rats fed sulphur amino acid-deficient diets. Can. J. Animal Sci. 58, 49-54.

Koster, H., Halsema, I.C.M. and Mulder, G.J. (1981). Factors that change the balance between sulfation and glucuronidation in isolated rat hepatocytes. Submitted.

Krijgsheld, K.R., Frankena, H., Scholtens, E., Zweens, J. and Mulder, G.J. (1979). Absorption, serum levels and urinary excretion of inorganic sulfate after oral administration of sodium sulfate in the conscious rat. Biochim. Biophys. Acta 586, 492-500.

Krijgsheld, K.R., Scholtens, E. and Mulder, G.J. (1980). Serum concentration of inorganic sulphate in mammals: species differences and circadian rhythm. Comp. Biochem. Physiol. 67A, 683-686.

Krijgsheld, K.R., Scholtens, E. and Mulder, G.J. (1981). An evaluation of methods to decrease the availability of inorganic sulfate for sulfate conjugation in the rat in vivo. Submitted.

Lakshmanan, F.L., Vaughan, D.A. and Barnes, R.E. (1978). Urinary inorganic sulfate excretion in human subjects fed purified amino acid mixtures with and without threonine. Fed. Proceed. 37, abstract 783.

Levinson, C. (1978). Chloride and sulfate transport in Ehrlich ascites tumor cells: evidence for a common mechanism. J. Cell. Physiol. 95, 23-32.

Levy, G. and Matsuzawa, T. (1967). Pharmacokinetics of salicylamide elimination in man. J. Pharmacol. exp. Therap. 156, 285-290.

Lindahl, U. and Höök, M. (1978). Glycosaminoglycans and their binding to biological macromolecules. Annu. Rev. Biochem. 47, 385-417.

Magdalou, J., Steimetz, D., Batt, A.M., Poullain, B., Siest, G. and Debry, G. (1979). The effect of dietary sulfur-containing amino acids on the activity of drug-metabolizing enzymes in rat liver microsomes. J. Nutr. 109, 864-871.

Moldeus, P., Andersson, B. and Gergely, V. (1979). Regu-   143
lation of glucuronidation and sulfate conjugation in
isolated hepatocytes. Drug Metabol. Disposit. <u>7</u>, 416-
419.
Mudge, G.H., Berndt, W.O. and Valtin, H. (1973). Tubular transport of urea, glucose, phosphate, uric acid
sulfate and thiosulfate. In <u>Handbook of Physiology</u>,
<u>Renal Physiology</u>, p. 587-652. Amer. Physiol. Soc.
Bethesda, MD.
Mulder, G.J. (1981a) Sulfation in vivo and in isolated
intact cell preparations. In <u>Sulfation of Drugs and</u>
<u>Related Compounds</u> (ed. G.J. Mulder), chapter 6, CRC
Press, Boca Raton, FL.
Mulder, G.J. (1981b). Sulfate availability in vivo. In
<u>Sulfation of Drugs and Related Compounds</u> (ed. G.J.
Mulder), chapter 3, CRC Press, Boca Raton, FL.
Mulder, G.J. and Keulemans, K. (1978). Metabolism of
inorganic sulphate in the isolated perfused rat liver. Biochem. J. <u>176</u>, 959-965.
Mulder, G.J. and Meerman, J.H.N. (1978). Glucuronidation and sulphation in vivo and in vitro: selective
inhibition of sulphation by drugs and deficiency of
inorganic sulphate. In <u>Conjugation Reactions in Drug</u>
<u>Biotransformation</u> (ed. A. Aitio), p. 389-387, Else-
vier Publ. Comp., Amsterdam.
Mulder, G.J. and Scholtens, E. (1978). The availability
of inorganic sulphate in blood for sulphate conjugation of drugs in rat liver in vivo. Biochem. J. <u>172</u>,
247-251.
Pang, K.S., Koster, H., Halsema, I.C.M. and Mulder, G.J.
(1981). Pharmacokinetics of sulfation and glucuronidation of harmol in the perfused rat liver. Submitted.
Rosenblum, W.I. (1968). Neuropathologic changes in a
case of sulfite oxidase deficiency. Neurology <u>18</u>,
1187-1192.
Roy, A.B. (1981). Sulfotransferases. In <u>Sulfation of</u>
<u>Drugs and Related Compounds</u> (ed. G.J. Mulder), chap-
ter 5, CRC Press, Boca Raton, FL.
Sabry, Z.I., Shadarevian, S.B., Cowan, J,W. and Campbell, J.A. (1965). Relationship of dietary intake of
sulphur amino-acids to urinary excretion of inorganic
sulphate in man. Nature <u>206</u>, 931-933
Schwarz, L.R. (1980). Studies with isolated hepatocytes: dependence of sulfation on cofactor supply.
Naun. Schm. Archs Pharmacol. <u>311</u>, abstract R27.
Schwarz, L.R. (1980). Modulation of sulfation and glucuronidation of 1-naphthol in isolated rat liver
cells. Arch. Toxicol. <u>44</u>, 137-145.

144   Shih, V.E., Abrams, I.F., Johnson, J.L., Carney, M., Mandell, R., Robb, R.M., Cloherty, J.P. and Rajago-palan, K.V. (1977). Sulfite oxidase deficiency. New Engl. J. Med. 297, 1022-1028.

Singer, T.P.(1975). Oxidative metabolism of cysteine and cystine in animal tissues. In Metabolic Path-ways, vol. VII, Metabolism of Sulfur Compounds (ed. D.M. Greenberg), 3rd edition, chapter 14, Academic Press, New York, N.Y.

Stipanuk, M.H. (1979). Effect of excess dietary methio-nine on the catabolism of cysteine in rats. J. Nutr. 109, 2126-2139.

Tateishi, N., Higashi, T., Naruse, A., Nakashima, K., Shiozaki, H. and Sakamoto, Y. (1977). Rat liver glu-tathione: a possible role as a reservoir of cysteine. J. Nutr. 107, 51-60.

Thornton, J.R. and English, P.B. (1977). Radiosulphate: its metabolism and use in measurement of extracellu-lar fluid volume in calves. Res. Veterin. Sci. 22, 298-302.

Weitering, J.G., Krijgsheld, K.R. and Mulder, G.J. (1979). The availability of inorganic sulphate as a rate limiting factor in the sulphate conjugation of xenobiotics in the rat? Biochem. Pharmacol. 28, 757-762.

White, C.J.B. (1979). Some characteristics of sulfate uptake into synaptosomes. Z. Naturforsch. C: Biosci. 34, 487-489.

Whiting, S.J. and Draper, H.H. (1980). The role of sul-fate in the calciuria of high protein diets in adult rats. J. Nutr. 110, 212-222.

Wong, K.P. and Yeo, T. (1979). Assay of adenosine 3'-phosphate 5'-sulphatophosphate in hepatic tissues. Biochem. J. 181, 107-110.

Zezulka, A.Y. and Calloway, D.H. (1976). Nitrogen re-tention in men fed isolated soybean protein supple-mented with L-methionine, D-methionine, N-acetyl-L-methionine, or inorganic sulfate. J. Nutr. 106, 1286-1291.

Phenolsulfotransferase Activity and Levels of Catecholamine Sulfates in Rat Brain Tissues

Nguyen T Buu and Otto Kuchel

Laboratory of the Autonomic Nervous System, Clinical Research Institute,
University of Montreal, Canada

INTRODUCTION

Sulfation constitutes a major pathway in the metabolism of substances containing phenolic hydroxyl groups but its physiological significance and its involvement in human diseases is still unclear. Thus while there was evidence that an increased sulfation of glucocorticoid accompanies elevated blood pressure in several types of experimental hypertension in rats (Singer et al, 1977) other report suggested a lack of sulfoconjugation of tyramine as a contributory factor in tyramine sensitive migraine (Youdim et al, 1971). Earlier Richter and MacIntosh (1941) had demonstrated that conjugation of adrenaline markedly reduced its pressor properties and proposed that conjugation is a mechanism of inactivation of the biological properties of this amine.

The physiological role of these sulfation may not be obvious but the chemical and physical properties of these esters differ considerably from those of the parent compounds. In particular, sulfation of the catecholamines (CA) converts these highly labile hormones into relatively stable products resistant to monoamine oxidase (MAO) (Renskers et al, 1980) and catechol-0-methyltransferase (COMT) degradations (Buu et al, 1981). Moreover conversion of an unionized group (e.g. free dopamine) into a fully ionized group such as dopamine 3-0-sulfate generally increases its permeability through tissue membrane (Jenner and Rose, 1973) and may facilitate its transport.

On the other hand the sulfate esters of CA may not be metabolically inert. Thus DA 3-0-sulfate has been shown to be metabolized in vivo in dogs and rats (Merits ,1976) and both DA 3-0-sulfate and DA 4-0-sulfate have been demonstrated to serve as substrates for dopamine β-hydroxylase (DβH) and be converted to free norepinephrine (NE) in vitro (Buu and Kuchel, 1979a, 1979b).

The enzyme responsible for the sulfoconjugation of the CA is phenolsulfotransferase (PST, EC. 2.8.2.1). Kinetic properties of its reactions have been reported with a partially purified enzyme (Banerjee & Roy, 1968: Wong, 1978) or enzymes purified to homogeneity (Sekura and Jacoby, 1979). Activity of PST has been detected in various tissues (Foldes and Meek, 1974) and in human blood platelets (Hart et al, 1979). Most recently, Anderson and Weinshilboum (1980) found a significant correlation between PST activity in human platelet and those in the renal cortex and jejunal mucosal, raising the possibility that platelet PST activity may serve as a marker of the degree of sulfate conjugation in man.

The role of PST in the regulation of CA levels in tissues has not yet been determined. It is not known for instance how much the PST activity in a tissue is indicative of the degree of conjugation of the CA in that same tissue. CA moreover do not exhibit the same affinity for the PST. Thus free DA showed the highest affinity for liver or platelet PST while free E showed the lowest (Meek & Neff, 1973; Kuchel et al, 1981). Whether the different affinities for the enzyme may reflect the relative levels of sulfate esters of each CA in tissues is not yet established.

The following study will attempt to deal with the above questions by comparing the levels of free and sulfoconjugated CA to levels of PST activity in different regions of rat brain.

METHODS

Sprague Dawley rats (150-200 g) were killed by decapitation. Their brain were quickly excised and dissected on ice according to the demarcations of Glowinski and Iversen (1966). Brain tissue were homogenized in 5 mM Tris HCl buffer, pH 7.5. For the measurement of the PST activity aliquots of the homogenate were centrifuged at 50,000 g at 4^0C for 1 hour. The clear supernatant was stored at -85^0C until analysis when it will be diluted accordingly in potassium phosphate buffer (5 mM, pH 7.5 with 0.0625% bovine serum albumine (BSA)). PST activity was measured essentially as described by Anderson and Weinshilboum (1980) except that the [^{35}S] PAPS (New England Nuclear, Mass., specific activity: 5 Ci/mmole) was diluted with non labelled PAPS (Pierce Chemical, Wis.) so that its specific activity became 1 Ci/mmole. Both 3-methoxy-4-hydroxyphenylethyleneglycol (MOPEG) and DA (30 mM) were used as substrate. Free and sulfoconjugated CA were measured as described in detail elsewhere (Buu et al, 1981). Essentially, free CA in brain tissues were adsorbed by alumina in Tris buffer at a pH 8.6, and were eluted from alumina with 0.1 N perchloric acid. CA sulfates, remaining in the fraction not adsorbed by alumina was passed through a Dowex column 50 X 8 (H+ form, 200-400 mesh) prepared in pasteur pipettes and eluted with water. CA were measured by the radioenzymatic assay of da Prada and Zürcher (1976) using [^{3}H] SAMe (New England Nuclear, Mass.) as a methyl donor and partially purified COMT (Axelrod and Tomchick, 1961) as the methylating enzyme. CA Sulfates were measured as above except that, in addition to COMT, a solution of sulfatase (Sigma, Mo) was added to the reaction mixture: CA sulfate was therefore hydrolyzed by sulfatase and the liberated free CA was then methylated by COMT. Such a hydrolysis by sulfatase had been shown to be quantitative.

RESULTS AND DISCUSSION

The presence of CA sulfate in brain has never been reported before although Meek and Neff (1973) found that injection of [^{35}S] sodium sulfate into rat brain was followed by the appearance of radiolabelled sulfated catecholamines or their sulfated metabolites. That CA sulfates are present in brain is compatible with the fact that PST activity has been detected (Foldes and Meek, 1974) in various regions of rat brain, most of them rich in free CA, particularly in free DA and NE. The difficulty of detecting CA sulfate in brain may due to the fact that their presence is masked by the much larger concentrations of their free form. Thus in the present method of analysis free CA were separated from the conjugated CA by an alumina adsorption. The results of measurements of PST activity, free and conjugated CA were summarized in the following table.

TABLE 1

Phenolsulfotransferase activity and concentrations of free

and sulfoconjugated CA in different regions of rat brain

Regions (number of rats)	PST*nmole/min mg protein	DA ng/g tissue free	DA ng/g tissue sulfate	E ng/g tissue free	E ng/g tissue sulfate	NE ng/g tissue free	NE ng/g tissue sulfate
Hypothalamus (10)	1.22×10^{-2}	112 ± 42	57 ± 15	69 ± 23	27 ± 6	1361 ± 205	37 ± 12
Striatum (10)	0.46×10^{-2}	5006 ± 540	48 ± 20	112 ± 19	33 ± 20	248 ± 33	20 ± 11
Hippocampus (8)	0.36×10^{-2}	83 ± 32	25 ± 7	36 ± 11	15 ± 13	299 ± 20	N.D.
Cerebellum (6)	N.D.	26 ± 8	N.D.	N.D.	N.D.	272 ± 19	N.D.
Cortex (6)	0.28×10^{-2}	777 ± 71	N.D.	20 ± 2	N.D.	290 ± 48	N.D.

N.D. = non detectable

*3-methoxy,4-hydroxyphenylethyleneglycol (MOPEG) (30 mM) was used as substrate

It shows that whereas PST activity was found in most regions of rat brain, except in the cerebellum, CA sulfates were detected only in the hypothalamus, striatum and in smaller quantities, in the hyppocampus, indicating that the presence of PST activity and of free CA substrates in a tissue does not necessarily lead to CA sulfate in that tissue. It is not known whether this is due to a lack of the cofactor PAPS within the tissue, or an inability of the tissue itself to retain CA sulfate which may have been formed. Those may also be the reasons why the striatum, with a large pool of free DA, has only a small concentration of DA sulfate. Levels of sulfate esters of CA do not seem to reflect their respective affinities for PST. Thus within the striatum the degree of sulfation is the highest for E and the lowest for DA contrary to what is expected from their respective affinities for PST. Surprisingly we found that PST activity in brain homogenates was almost twice higher when free DA was used as substrate instead of MOPEG. This may be due to the higher affinity of DA for PST and points out perhaps to the difference of enzyme activity in vitro and in vivo.

The role of sulfation of CA in brain, as well as that of CA sulfates, is not known. PST is not a specific enzyme since it can accept a large variety of substrates: monophenolic compounds, catechols, catecholamines and their numerous metabolites (Barnarjee and Roy, 1968). Its activity is widely distributed in both neurogenic and non-neurogenic tissues. The low activity of the enzyme (Table 2) in the richly innervated tissues such as the heart or spleen together with the very small concentration of CA sulfates in these tissues (unpublished results) speak against the role of sulfation as a major inactivation pathway of CA.

TABLE 2

Ratios of relative PST activities

in different rat tissues

Liver	1.0
Kidney	0.014
Spleen	0.005
Heart	0.003

Sulfoconjugation of CA on the other hand made them resistant to degradations by MAO and COMT and extend considerably their half-life in vivo. The CA sulfates are now capable of forming internal salt and pass more freely through cell membrane. These properties are suggestive of the role of CA sulfates as transport and storage form for the CA, at least in the periphery. In brain the small presence of CA sulfate, amid much larger quantities of their free form,is surprising. It is possible that CA sulfate, unlike the free CA, can easily cross the blood brain barrier into the periphery. Meek and Neff (1972) for instance have demonstrated that the sulfate ester of MOPEG can pass readily the blood brain barrier whereas free MOPEG cannot. If the same thing can be demonstrated for CA sulfate, sulfation may constitute a means of transporting the excess of free CA from brain to the periphery. It is also possible that brain tissues, like other richly innervated tissues in the periphery, are not capable of retaining CA sulfate, which then pass quickly into the cerebrospinal fluid.

In conclusion the results from this study indicated that measurement of PST activity alone does not provide any information as to what is the degree of conjugation of CA in the tissue. This,in turn,does not depend on the availability of substrate in the tissues. Sulfation of CA in brain may constitute a means of transporting CA from brain to the periphery where CA sulfate may serve as a storage form and possibly precursors for free CA.

This work was supported by a grant from the Medical Research Council of Canada to a multidisciplinary research group in hypertension.

REFERENCES

Anderson R.J. and Weinshilboum R. (1980). Phenolsulpho-transferase in Human Tissue: Radiochemical Enzymatic Assay and Biochemical Properties. Clin. Chim. Acta 103:79-90.

Anderson R., Weinshilboum R., Philips S. and Broughton D. (1980). Human Platelet Phenolsulphotransferase (PST) Activity: Correlation with PST Activity in the Kidney and Gut. Pharmacologist 22(3): p. 301.

Axelrod J. and Tomchick R. (1958). Enzymatic 0-methylation of Epinephrine and other Catechols. J. Biol. Chem. 233:702-705.

Barnerjee R.K. and Roy A.B. (1968). Kinetic Studies of the Phenolsulphotransferase reaction. Biochim. Biophys. Acta 151:573-586.

Buu N.T. and Kuchel O. (1979a). The Direct Conversion of Dopamine 3-0-sulfate to Norepinephrine by Dopamine-β-Hydroxylase. Life Sci. 24:783-790.

Buu N.T. and Kuchel O. (1979b).Dopamine 4-0-sulfate: A Possible Precursor of Free Norepinephrine. Can. J. Biochem. 57:1159-1162.

Buu N.T., Duhaime J., Savard C., Truong L. and Kuchel O. (1981). Presence of Conjugated Catecholamines in Rat Brain: A New Method of Analysis of Catecholamine Sulfates. J. Neurochem. (in press).

Da Prada M. and Zürcher G. (1976). Simultaneous Radioenzymatic Determination of Plasma and Tissue Adrenaline, Noradrenaline and Dopamine within the Femtomole Range. Life Sci. 19:1161-1174.

Foldes A. and Meek J.L. (1974). Occurence and Localization of Brain Phenolsulphotransferase. J. Neurochem. 23:303-307.

Glowinski J. and Iversen L.L. (1966). Regional Studies of Catecholamines in the Rat Brain. 1. The Disposition of [^{3}H] Norepinephrine, [^{3}H] Dopamine and [^{3}H] Dopa in Various Regions of the Brain. J. Neurochem. 13:655-669.

Hart R.F., Renskers K.J., Nelson E.B., Roth J.A. (1979). Localization and Characterization of Phenolsulfotransferase in Human Platelets. Life Sci. 24:125-130.

Jenner W.N. and Rose F.A. (1973). Studies on the Sulphation of 3-4-dihydroxyphenylethylamine (Dopamine) and Related Compounds by Rat Tissues. Biochem. J. 135:109-114.

Kuchel O., Buu N.T., Hamet P., Larochelle P. and Bourque M. (1981). Catecholamine Sulfates and Platelet Phenolsulfotransferase (PST) Activity in Human Hypertension. In Phenolsulfotransferase, Usdin E. & Sandler M. (eds) Macmillan Publishers Ltd. England, in press.

Meek J.L. and Neff N.H. (1972). Acidic and Neutral Metabolites of Norepinephrine: Their Metabolism and Transport from Brain. J. Pharmacol. Exp. Ther. 181:457-462.

Meek J.L. and Neff N.H. (1973). Biogenic Amines and their Metabolites as Substrates for Phenol Sulphotransferase (EC 2.8. 2.1.) of Brain and Liver. J. Neurochem. 21:1-9.

Merits I. (1976). Formation and Metabolism of [14C] Dopamine 3-0-sulfate in Dog, Rat and Guinea Pig. Biochem. Pharmacol. 25:829-833.

Renskers K.J., Foer K.D., and Roth J.A. (1980). Sulfation of Dopamine by Human Brain Phenol Sulfotransferase. J. Neurochem. 34:1362-1368.

Richter D. and MacIntosh F.C. (1941). Adrenaline Ester. Am. J. Physiol. 135:1-5.

Sekura R.D. and Jacoby W.B. (1979). Phenolsulfotransferase. J. Biol. Chem. 254:5658-5663.

Singer S.S., Hess E. and Sylvester S. (1977). Hepatic Cortisol Sulfotransferase Activity in Several Types of Experimental Hypertension in Male Rats. Biochem. Pharmacol. 26:1033-1038.

Wong K.P. (1978). Measurement and Kinetic Study of the Formation of Adrenaline Sulfate in vitro. Biochem. J. 174:777-782.

Youdim M.B.H., Carter S.B., Sandler M., Hanington E. and Wilkinson M. (1971). Conjugation Defect in Tyramine Sensitive Migraine. Nature 230:127-128.

In Vivo and in Vitro Efflux of Conjugated Amines into Superfusates from Discrete Areas of the Central Nervous System of Cats and Rats

Gertrude M Tyce[1,2], Donna L Hammond[3], Duane K Rorie[4], and Tony L Yaksh[3]

Departments of Physiology[1], Biochemistry[2], Neurosurgical[3] Research, and [4]Anesthesiology, Mayo Foundation, Rochester, Minnesota, 55901 USA

INTRODUCTION

We have previously found (Tyce et al, 1980) considerable amounts of an acid-hydrolyzable conjugate of dopamine (DA), presumably DA-sulfate, in the cerebrospinal fluid (CSF) of dogs following intravenous administration of L-3,4-dihydroxyphenylalanine (L-DOPA). The amounts of this DA conjugate present in the CSF were decreased when α-methyldopahydrazine, which inhibits decarboxylation of L-DOPA in peripheral tissues, was administered in addition to L-DOPA. Since this treatment could be expected to preferentially increase central stores of DOPA and DA (Melamed et al, 1980), our data suggested that the DA conjugate present in the CSF of dogs had been formed in the periphery, and transported into the brain and CSF. Amine sulfates form internal salts carrying little electrical charge at physiological pH and might therefore penetrate biological membranes more readily than the parent amines which are not easily transported across biological membranes (Jenner & Rose, 1973).

In the present studies we have shown that conjugated DA is, in fact, released from the central nervous system. The release of conjugated DA was studied in vivo using ventriculo-cisternal perfusions of cats which had been pretreated with N-methyl-N-propargyl-3-(2,4-dichlorophenoxy)propylamine hydrochloride (clorgyline, May & Baker), an inhibitor of monoamine oxidase (MAO). The release of conjugated DA was also examined in vitro using slices of striatum obtained from rats. In addition, the release of a conjugate of 5-hydroxytryptamine (5HT, serotonin) was demonstrated using in vivo superfusions of the spinal cord in rats.

METHODS

Ventriculo-cisternal Perfusions in the Cat

Mongrel cats 2-4 kg, of either sex, were anesthetized with chloralose-urethane and endotracheal tubes inserted. Cannulae were inserted in a carotid artery for measurement of arterial blood pressure, and in an antecubital vein for the administration of clorgyline (16.0 mg/kg). After placement of the cat's head in a stereotaxic frame, a burr hole was drilled 2 mm posterior to bregma and 1 mm lateral to the midline suture of the skull. An 18 gauge stainless steel inflow cannula, which was attached to fluid-filled tubing, was then slowly advanced through the burr hole into the brain until a rapid drop in the fluid level of the tubing occurred. This rapid inflow of fluid was indicative of penetration of the lateral ventricle by the cannula. The cannula was then connected by PE-90 tubing to an inflow syringe containing artificial CSF (Yaksh & Tyce, 1980) and mounted on a Harvard infusion pump. The musculature overlying the atlantooccipital membrane was dissected away and a discrete puncture was made in the membrane. A short length of PE-90 tubing, which served as the outflow cannula, was inserted through the membrane such that it lay within the cisterna magna. Outflow was achieved by maintaining the end of the outflow catheter approximately 10 cm ventral to the cisternal magna. Thus, artificial CSF was infused into the lateral ventricle at a nominal rate of 0.1 ml/min and the outflow was collected passively from the cisternal magna in iced sampling tubes. Two-ml volumes of clear perfusate were routinely collected per 20 min time interval. The cats were perfused for 20 min prior to the collection of the first sample; this initial washout was discarded. Thereafter, samples of perfusate were collected continuously. The first sample was collected for 60 min; samples two through six were each collected during the five successive 20-min intervals (The first sample was divided into three aliquots and was used to monitor recoveries of added standard DA or DA-sulfate). The first and second samples of perfusate were collected under basal conditions, the third and fourth samples during the addition of tyramine (5×10^{-4}M) to the perfusate to evoke DA release, and the fifth and sixth samples after the replacement of tyramine-containing CSF with drug-free CSF.

Superfusions of Slices of Striatum from Rat Brains

Male Sprague-Dawley rats weighing 250-300 gm were killed by decapitation. The skulls were opened and the brain removed rapidly. Slices of striatum were cut (3 mm in diameter, 0.5 cm thick) and transferred to a superfusion chamber containing warmed, oxygenated Krebs-Ringer bicarbonate solution containing glucose (11 mM).

For each experiment eight slices from the brains of two animals were combined for study of release. The mean weight of the tissue was 71.3 $\pm$ 4.0 (S.D.) mg. The superfusion chamber consisted of a polypropylene cylinder 8.5 mm in diameter and 6 cm in length. The cylinder was converted into a closed chamber by plungers through which inlet and outlet tubes were fitted. The plungers were fitted into each end of the chamber and could be adjusted toward a central point in the cylinder to produce a chamber volume of 1 ml which was used throughout the experiments. Krebs-Ringer bicarbonate solution containing glucose, maintained at 37°C and aerated continuously with 5% carbon dioxide in oxygen in a reservoir container, was pumped continuously through the chamber at a flow rate of 0.5 ml/min by a roller pump (Extracorporeal, Model 911). Sodium metabisulfite was added to the Krebs-Ringer as an antioxidant (final concentration 100 mg/L). The temperature of the chamber was maintained constant by circulating 37° water around the chamber exterior. The slices were superfused for 45 min before experimentation began. After the experiments were started, the Krebs-Ringer solution leaving the superfusion chamber was collected in 10-min intervals (5 ml).

The first sample of superfusate was collected under basal conditions. Tyramine (5 x 10^{-4} M) was then added to the superfusate and two samples of superfusate were collected. Krebs-Ringer solution was then re-introduced as the superfusing fluid and an additional two samples of superfusate were collected.

In some experiments rats were treated with clorgyline (1 mg/kg, i.v.) 1 h before they were killed and slices of striatum removed for superfusion. This dose of clorgyline has been shown to cause 90% inhibition of MAO in rat brain between 1 and 4 h after administration (Yang & Neff, 1974). In other experiments clorgyline (final concentration 8 x 10^{-5} M) was present in the superfusate throughout the superfusion.

In Vivo Superfusion of the Rat Spinal Cord
Sprague-Dawley rats (275-325 g) were prepared under chloralose-urethane anesthesia for superfusion of the spinal cord subarachnoid space. PE-10 tubing was inserted through a transverse slit in the atlantooccipital membrane and threaded caudally in the spinal subarachnoid space for a distance of 10 cm. This catheter was connected to an inflow (push) syringe containing artificial CSF. A 20-gauge cannula was positioned under the atlantooccipital membrane in the cisternal magna and connected to the outflow (pull) syringe. These syringes were mounted on a Harvard infusion-withdrawal pump. Thus, artificial CSF was delivered at 0.1 ml/min to the caudal intrathecal space, and after superfusing the entire spinal cord, was collected at the

level of the cisternal magna in sampling tubes. Details of the
procedure are given elsewhere (Yaksh & Tyce, 1978).

Superfusate (4 ml) was collected under basal conditions for
40 min. We have shown previously (Hammond et al, 1980) that
the basal efflux of 5HT from rat spinal cord into superfusate
does not change significantly during this time. Additional
superfusate (4 ml; 40 min) was collected subsequently during the
addition of 10^{-3}M DL-p-chloroamphetamine to the superfusate.
This agent has been shown to evoke the release of 5HT in vitro
(Fuller & Molloy, 1974), and in this preparation in vivo
(Hammond et al, 1980).

Superfusions of the spinal cord were performed in control
rats and in rats which had been treated both 18 and 1 hr prior
to superfusion with clorgyline (8 mg/kg, i.m.). This dose regi-
men had been shown to produce at least 75% inhibition of MAO
(Johnston, 1980).

Measurement of Free and Conjugated Dopamine in Superfusates
Superfusates were collected into tubes standing in iced
water and containing 10 μl of 5% sodium metabisulfite for each
ml of superfusate to be collected. Free and conjugated DA were
first separated by alumina chromatography: Aliquots (2 to 5 ml)
of superfusate were added to tubes containing 1 ml of 2 M Tris-
HCl buffer, pH 8.6 containing 5% disodium EDTA, and 250 mg of
washed alumina. After shaking for 5 min the tubes were centrifuged and the supernate which contained conjugated DA was saved.
The alumina was washed twice with 5 ml of water and DA was eluted
from the alumina with 6 ml of 0.05 N perchloric acid. To an
aliquot (5.6 ml) of this eluate was added 0.1 ml of 5% disodium
EDTA and 0.1 ml of 0.1% ascorbic acid. The mixture was adjusted
to pH 6.0 to 6.2 with sodium carbonate (0.2 N) and applied to
columns (3 mm diameter; 3.0 cm height) containing Amberlite CG
50, 100 to 200 mesh. After washing with 5 ml of water, DA was
eluted from the resin with 1 ml of 0.67 M boric acid. When
maximum sensitivity was needed elution was with 5 ml of 0.5 N
acetic acid. After the addition of 25 μl of 10 mM disodium EDTA
the acetic acid eluate was lyophilized. The residue was
dissolved in 200 μl of 0.67 M boric acid immediately before
quantitation of DA.

Conjugated dopamine, present in the supernatant solution
after extraction of the superfusates with alumina, was hydrolyzed
by boiling in strong acid. To 3 to 5 ml aliquots of this supernate was added 0.2 ml of 5% sodium metabisulfite and 0.3 ml of
1% dithiothreitol. The mixture was adjusted to 0.8 M with 8 M
perchloric acid, and heated for 45 min at 100°. After cooling
in iced water for 15 min, the released DA was separated in the
same manner as the free DA.

Quantitation of DA was by liquid chromatography with electrochemical detection (LCEC). Separation of DA was achieved using a µBondapak C_{18} column and a mobile phase of 0.07 M phosphate buffer at pH 4.7, 0.5 mM heptanesulfonate, 5% methanol and 0.2 mM disodium EDTA. The electrochemical detector was maintained at 0.6 V versus a silver-silver chloride cell.

The mean recovery of 10 ng of DA added to 5 ml of superfusate before the preliminary two-column extraction procedure was $75.2 \pm 2.1\%$ (SEM, n = 18). The mean recovery of 10 ng of DA added to the alumina supernatant fraction before acid boiling and the two column procedure was $68.5 \pm 3.2\%$ (n = 15). The data have been corrected for these average recoveries. The interference from DA in the DA-conjugate determination was $0.7 \pm 0.3\%$ (n = 17).

Measurement of Free and Total (Free Plus Conjugated) 5HT and 5HIAA in Superfusates

Superfusate was collected into iced tubes containing 100 µl 4% cysteine as an antioxidant for each ml of superfusate to be collected. 5HT and 5HIAA in the superfusates were separated by cation exchange and gel chromatography before quantification by LCEC. Each sample of superfusate collected under basal and evoked-release conditions was split into two 2-ml fractions. One fraction was used to determine the amounts of free 5HT present. The remaining fraction was used to determine the amounts of total (free plus conjugated) 5HT present. For determination of free 5HT, the fraction was adjusted to pH 6, and applied to columns (4 mm diameter and 2.3 cm height) of Amberlite CG 50 (200-400 mesh) which had been equilibrated previously in 0.2 M phosphate buffer pH 6.1 containing disodium EDTA (0.1%). The resin was washed with 3 ml of 0.02 M phosphate buffer pH 6.1, and 3 ml water, and 5HT was eluted with 3 ml of a mixture of 1N formic acid and 1N HCl (85:15, v/v). Disodium EDTA (25 µl of a 10 mM solution) and 5 µl of a 1% cysteine solution were added to this eluate and it was then lyophilized, and kept frozen at −40° until quantification. The residue was dissolved in 500 µl of 0.05 M acetic acid immediately before quantification.

5HIAA was separated from aliquots (8 ml) of the combined effluents and washes from the CG 50 column chromatographically using a column (5 mm diameter) containing 0.5 g of Sephadex G10 which had been swollen for 1 h in 0.1 M HCl. After three washes with 2 ml of 0.1 M HCl, and one wash with 3 ml of water, 5HIAA was eluted from the Sephadex with 2.5 ml of 0.005 M phosphate buffer, pH 8.5, into a tube containing 25 µl of 10 mM disodium EDTA and 5 µl of 1% cysteine.

For measurement of total 5HT, 7.5 M perchloric acid (1/10

volume) was added to the remaining fraction of superfusate, and the mixture was heated for 30 min in a boiling water bath. The tubes were cooled in iced water for 15 min, and perchlorates were removed from the extracts by adjustment to pH 5 to 5.5 with 5 M KOH and 1 M KOH. After centrifugation, aliquots of the clear supernate were applied to CG 50 columns, and total 5 HT was separated by the same procedure used for separation of free 5HT. 5HIAA was destroyed by boiling in strong acid so this method could not be used for detection or measurement of conjugates of 5HIAA.

Quantitation of 5HT and 5HIAA in their respective fractions was by LCEC. A reversed phase separation, comprising a Biophase (Bioanalytical Systems) column, and a mobile phase of 0.07 M phosphate at pH 4.7, 0.2 mM disodium EDTA, and 12% methanol, was used. The electrochemical detector was maintained at 0.6 V versus a silver/silver chloride electrode.

The mean recovery of 10 ng of 5HT added to superfusate before the preliminary extraction was $84.2 \pm 3.5\%$ (n = 12). This was not different from the recovery of 10 ng 5HT added to superfusate before acid boiling and column extraction ($88.5 \pm 3.8\%$ (n = 11). The recovery of 40 ng 5HIAA added to superfusate was $85.4 \pm 3.1\%$ (n = 11).

RESULTS

In Vivo Release of Free and Conjugated DA into Ventriculo-cisternal Perfusates in Cats Treated with Clorgyline

The basal release of free and conjugated DA into ventriculo-cisternal perfusates of clorgyline-treated cats was not consistently demonstrable (Table 1, Figure 1). Amounts in excess of 0.2 to 0.3 ng per 2 ml of perfusate collected over a 20 min interval would have been detected with this method. Free DA and acid-hydrolyzable DA were released by the addition of tyramine to the perfusate; the amounts of conjugated DA represented from 8 to 12% of the total. When tyramine-containing perfusate was replaced with drug-free perfusate, the efflux of both free and conjugated DA decreased sharply.

In Vitro Release of Free and Conjugated DA from Slices of Striatum from Rats

Free but not conjugated DA was detected in superfusate of slices of striatum from control rats collected under basal conditions (Table 2). The sensitivity of the method was such that if overflow of conjugated DA occurred, the amounts released must have been less than 0.2 to 0.3 ng per 100 mg slices per 10 min. Addition of tyramine to the superfusate evoked the release of considerable amounts of free and conjugated DA, the amounts of

158

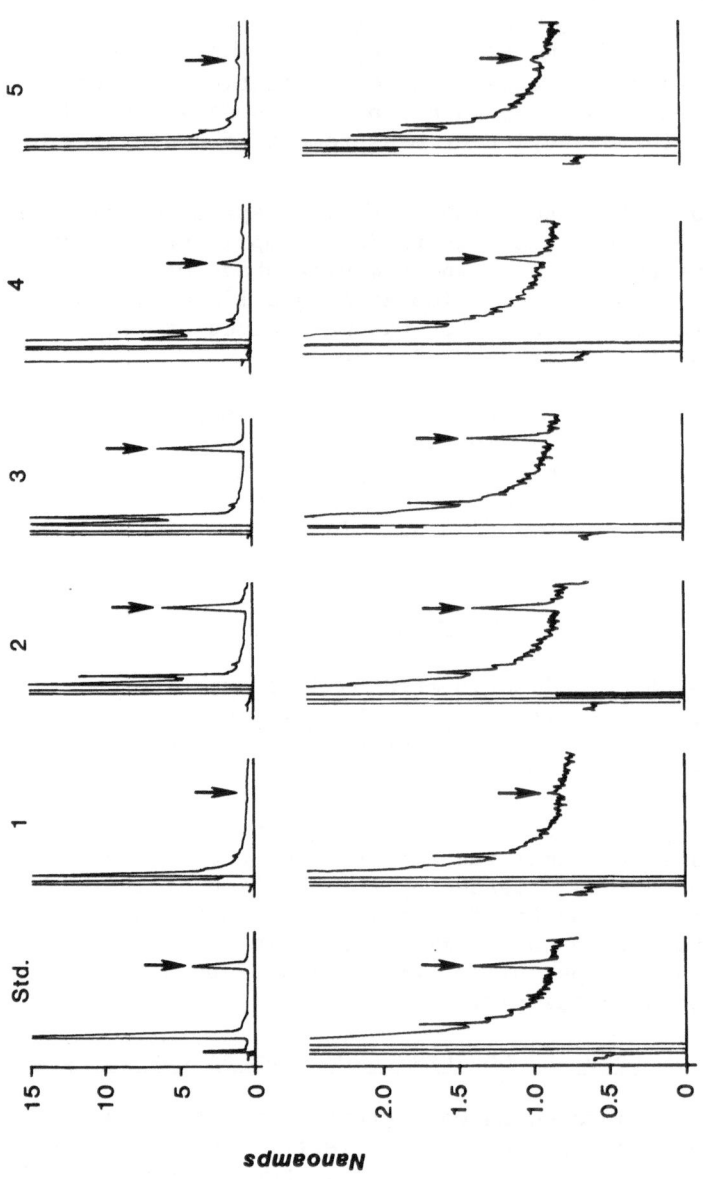

Figure 1. Chromatograms of extracts of ventriculo-cisternal perfusates in cats. Free DA (upper rank), and DA freed by acid hydrolysis from the same samples after removal of the free DA by adsorption on alumina (lower rank), are shown. DA eluted after 10 min.
Std: 250 pg DA, upper rank; 40 pg, lower rank.
Sample one was collected under basal conditions, samples two and three when release of DA was evoked by tyramine (5 x 10^{-4} M), and samples four and five in the immediate post-treatment period.

TABLE 1. Free and conjugated dopamine in ventriculo-cisternal[1]
perfusates of cats treated with clorgyline[1]

Time period	DA released; ng/20 min	
	Free	Conjugated
1.[2] (Basal release)	0.2[3]	0.6[3]
2. (Basal release)	0.4[3]	0.7[3]
3. (Tyramine[4]-evoked release)	7.7±1.1	1.1±0.2
4. (Tyramine-evoked release)	14.1±3.6	1.5±0.2
5. (Post tyramine)	10.1±3.5	0.9±0.2
6. (Post tyramine)	2.8,0.5[3]	0.2[3]

[1] Data are given as mean ± SEM of four experiments.
[2] Sample one was collected over 60 min and samples two through
six were each collected over 20 min. Clorgyline (16 mg/kg,i.v.)
was administered 60 min before collection of sample one.
[3] Free and conjugated dopamine were not consistently measurable
in samples one, two and six; the values shown for these time
periods are single values in which the amounts of amine could
be measured.
[4] The tyramine concentration was 5×10^{-4} M.

conjugated DA representing about 10% of the total. During the
post stimulation period the amounts of free and conjugated DA in
superfusate decreased.

When clorgyline (8×10^{-5} M) was present in the superfusate,
increased amounts of free DA were released by tyramine (Table 2).
The amounts of conjugated DA also appeared to be increased, but
the differences were only significant in the final sample after
stimulation. In rats pretreated systemically with clorgyline,
the release of free DA was greater than that in control,
untreated rats, and the release of conjugated DA was greater
under basal conditions, in the first sample collected during
tyramine infusion, and in the final sample after stimulation.

160

TABLE 2. Release of free and conjugated DA from slices of striatum from rat brain[1]

Time period	Free DA			Conjugated DA		
	Control	Clorgyline (8×10^{-5} M)	Clorgyline pretreatment[2]	Control	Clorgyline (8×10^{-5} M)	Clorgyline pretreatment
1. (Basal release)	6.9±1.5	5.5± 1.6	3.5± 0.5	--	--	1.2±0.3[3]
2. (Tyramine[4]-evoked release)	60.2±9.6	88.8± 6.9[3]	124.5±11.2[3]	7.2±0.7	11.1±2.5	16.2±3.7[3]
3. (Tyramine-evoked release)	62.2±7.3	93.0±10.0[3]	124.4±13.0[3]	9.1±1.3	13.0±3.8	14.7±2.9
4. (Post tyramine)	21.1±3.4	39.8± 6.1[3]	46.2± 7.2[3]	2.6±0.6	3.6±1.3	6.9±2.2
5. (Post tyramine)	8.5±1.3	17.3± 4.1	15.2± 2.9	--	0.3±0.1[3]	1.0±0.6[3]

[1] data as mean (± SEM, n = 5) ng DA released per 100 mg tissue per 10 min.
[2] Clorgyline (1 mg/kg, i.v.) was given 1 h before the rats were killed and the slices were cut.
[3] significant difference (p < .05) from controls.
[4] the concentration of tyramine was 5×10^{-4} M.

TABLE 3. Release of 5HT and metabolites into superfusates of rat spinal cord[1]

Experimental Conditions	ng released per 20 min		
	5HT	5HT-conj	5HIAA
Control Rats:			
Basal release	0.50±0.05	0.10±0.02	15.6±1.0
PCA[2]-evoked release	1.82±0.10[3]	0.15±0.10	16.1±1.6
Clorgyline[4]-treated Rats:			
Basal release	0.81±0.10[5]	0.15±0.06	8.2±0.6[5]
PCA-evoked release	2.87±0.46[3,5]	0.20±0.08	6.7±0.5[5]

[1] data are given as mean (± SEM) of 5 determinations in control rats and 6 determinations in clorgyline-treated rats.
[2] PCA: p-chloroamphetamine, 10^{-3} M.
[3] significant difference ($p < .05$) from basal release.
[4] Clorgyline (8 mg/kg i.m.) was given 18 h and 1 hr prior to superfusion.
[5] significant difference ($p < .05$) from control rats.

<u>In Vivo Release of Free and Conjugated 5HT from Rat Spinal Cords</u>

Free 5HT was detected in spinal superfusates collected under basal conditions (Table 3). In all samples, boiling in strong acid increased the amounts of 5HT in superfusate, the increase being between 20 and 25% of the total 5HT. The amounts of 5HIAA in superfusate greatly exceeded the amounts of either free or acid-hydrolyzable 5HT.

In control rats the addition of p-chloroamphetamine to the superfusate evoked the release of 5HT, but not of acid-hydrolyzable 5HT, nor of 5HIAA.

The amounts of free 5HT in superfusates of spinal cord were greater under basal conditions in clorgyline-treated rats than in controls, and the amounts of 5HIAA were less. There were no

differences in the amounts of conjugated 5HT present in superfusates collected under basal conditions between clorgyline-
treated and control rats. p-Chloroamphetamine evoked the release
of greater amounts of 5HT from clorgyline-treated rats than from
controls. p-Chloroamphetamine did not evoke release of 5HIAA
nor of conjugated 5HT from clorgyline-treated rats.

DISCUSSION

The release of small amounts of an acid-hydrolyzable conjugate of DA from the central nervous system has been demonstrated under in vivo and in vitro conditions in the present
experiments. It was not shown that the conjugate that was
released was, in fact, DA-sulfate. However, it seemed likely
that the conjugate was a sulfate because it was hydrolyzed by
acid. Many catecholamine metabolites form glucuronide conjugates,
but these are acid stable (Weil-Malherbe, 1971). The conjugate
was not N-acetyldopamine, since this compound was tested and
found not to interfere in the determination of free and conjugated DA.

DA-sulfate has recently been found in brain by Buu et al
(1980). In rat striatum the ratio of conjugated DA/free DA was
1:100. This was much less than the corresponding ratio (1:10)
found in superfusates of slices of striatum in the present experiments. This suggests that when conjugated DA is formed it
leaves brain rapidly.

An acid-hydrolyzable conjugate of 5HT was also detected in
superfusates of rat spinal cord. The amounts of the conjugate
were not increased significantly when the release of free 5HT
was evoked by p-chloroamphetamine.

Clorgyline did not have dramatic effects on the amounts of
conjugated amines detected in the present studies. Thus, clorgyline pretreatment had no signficant effects on the release of
the acid-hydrolyzable conjugate of 5HT, either under basal conditions or during p-chloroamphetamine-induced release of 5HT.
In addition, when slices of striatum were taken from brains of
rats which had been pretreated with clorgyline, increased
amounts of conjugated DA were present in superfusate, but the
increases were modest.

The decreases in 5HIAA in superfusate which were apparent in
superfusates of rat spinal cord when the animals had been pretreated with clorgyline were not compensated for by increases in
acid-hydrolyzable or free 5HT. This might suggest that one of
the actions of clorgyline was to inhibit release of 5HT. However,

it is likely that the egress of 5HIAA and of conjugated 5HT
from CSF into blood occur at different rates. Further experiments studying release of 5-hydroxyindoles from brain slices
are needed because differences in rates of clearance from CSF to
blood are avoided with this technic. Such studies should clarify whether the decreases in the amounts of 5HIAA in superfusate
are compensated for by increases in conjugated 5HT when MAO is
inhibited, or whether clorgyline, in fact, inhibits release of
amines.

REFERENCES

Buu, N. T., Duhaime, J., Savard, C., Truong, L. and Kuchel,
O. (1980). J. Neurochem. in press.

Fuller, R. W. and Malloy, B. B. (1974). Adv. in Biochem.
Psychopharmacol., 10, 195-205.

Hammond, D. L., Tyce, G. M. and Yaksh, T. L. (1980). The
Pharmacologist, 22(3), 219.

Jenner, W. N. and Rose, F. A. (1973). Biochem. J., 135,
109-114.

Johnston, J. P. (1968). Biochem. Pharmacol., 17, 1285-1297.

Melamed, E., Hefti, F. and Wurtman, R. J. (1980). Brain
Res., 198, 244-248.

Tyce, G. M., Sharpless, N. S., Kerr, F. W. L. and Muenter,
M. D. (1980). J. Neurochem., 34, 210-212.

Weil-Malherbe, H. (1971). In Methods of Biochemical Analysis, Suppl. Vol. (ed. D. Glick), pp. 119-152. Interscience Publishers, New York.

Yaksh, T. L. and Tyce, G. M. (1980). Brain Res., 192, 133-
146.

Yang, H.-Y. T. and Neff, N. H. (1973). J. Pharmacol. Exp.
Ther., 187, 365-373.

Conjugated Dopamine in Human Ventricular Fluid

Nansie S Sharpless[1], Gertrude M Tyce[2], Leon J Thal[3],
Joseph M Waltz[4], Kamran Tabaddor[5] and Leslie I Wolfson[3]

Departments of Psychiatry[1], Neurology[3], and Neurosurgery[5]
Albert Einstein College of Medicine, Bronx, New York 10461 USA
Department of Biochemistry and Physiology[2],
Mayo Clinic and Mayo Foundation, Rochester, Minnesota 55901 USA
Department of Neurosurgery[4],
St Barnabas Hospital, Bronx, New York 10457 USA

INTRODUCTION

Measurement of cerebrospinal fluid (CSF) constituents is
the most direct approach available for study of cerebral neurotransmitter release and turnover in man. Numerous studies have
demonstrated that neurotransmitter amines and their metabolites
in CSF derive from the central nervous system and not from
peripheral tissues (Pletscher et al., 1967; Moir et al., 1970;
Garelis et al., 1974). CSF amine metabolite concentrations have
also been shown to reflect neuronal aminergic activity in
adjacent brain structures (Portig and Vogt, 1969; Moir et al.,
1970; Papeschi et al., 1971). Because sensitive fluorometric
methods have been available for measuring homovanillic acid
(HVA), the major metabolite of dopamine, there have been numerous
studies of lumbar CSF concentrations of HVA in patients with
psychiatric or neurological disorders (for reviews, see Wood,
1980). Due to the presence of mediated transport systems for
removal of organic acids from CSF (Wolfson et al., 1974), there
have been questions about the extent to which lumbar CSF acid
metabolite concentrations reflect cerebral aminergic activity.
The nigrostriatal dopaminergic pathway, which is important in
regulation of movement, terminates in the caudate nuclei directly
adjacent to the lateral ventricles. We have been studying
dopamine and its metabolites in samples of ventricular fluid
obtained during ventriculography prior to thalamic surgery in
patients with movement disorders (Tabaddor et al., 1978 a,b).
In patients with Parkinson's disease, the well documented
decrease in dopamine and HVA concentrations in the caudate
nucleus was accompanied by decreased HVA concentrations in
ventricular CSF (Papeschi et al., 1972; Tabaddor et al., 1978a).

164

Lesser decreases in ventricular CSF HVA levels were seen in patients with clinical evidence of diffuse cerebral disease. In patients with adult-onset dystonia, ventricular fluid HVA levels were significantly lower than in patients with childhood-onset dystonia, suggesting that there might be diminished dopaminergic activity in adult-onset dystonia (Tabaddor et al., 1978b). These observations led us to expand our studies to include measurements of dopamine itself.

Neurotransmitter amines released from peripheral or central nerve endings are generally inactivated by reuptake or metabolism (Glowinski and Axelrod, 1966). However, as has been demonstrated for norepinephrine, some amine does escape into CSF where its concentration appears to reflect neuronal activity in adjacent nervous tissue (Ziegler et al., 1976; 1977). We reasoned that if dopamine, which is not normally measureable in human lumbar fluid (Peuler and Johnson, 1977; Levin and Hubschmann, 1980), were released into CSF, it would be detectable in human ventricular fluid. We found that, while dopamine is present in human ventricular fluid, it exists there predominantly in the form of an acid hydrolyzable conjugate. This observation has important implications in relation to the mechanisms of neuronal dopamine inactivation and removal from brain and CSF.

METHODS

Samples of ventricular fluid were obtained at the time of ventriculostomy in patients with movement disorders. Patients who were taking drugs (other than L-DOPA) known to affect monoamine metabolism were excluded. All of the fluids were free of visible erythrocyte contamination. They were centrifuged, immediately frozen, and stored at $-30^{\circ}C$ until assay.

Free dopamine, free norepinephrine, and dopamine released by hydrolysis were measured with two different methods: radioenzymatic assay and high-performance liquid chromatography (HPLC) with electrochemical detection. Full details of these methods have been described elsewhere (Sharpless et al., 1980a,b) and are briefly outlined here.

In the radioenzymatic assay, free catechols were radiolabelled by catechol-O-methyltransferase (COMT) using (^{3}H-methyl)-S-adenosyl-L-methionine as methyl donor. The labelled products were extracted into ether, returned to acid, and then separated by thin-layer chromatography on Whatman K6DF plates developed in chloroform:ethanol:70% ethylamine (80:17:9). Conjugates were hydrolyzed by lyophilization in dilute $HClO_4$ (Buu and Kuchel, 1977) and the released dopamine was measured radioenzymatically. Recovery of dopamine liberated from authentic dopamine sulfate (synthesized in our laboratory using the method of Jenner and Rose, 1973) was 85% from dopamine-3-0-sulfate and 97% from dopamine-4-0-sulfate with this method of hydrolysis.

For HPLC, free catechols were removed from an aliquot of CSF by adsorption onto alumina and then eluted from the alumina with acetic acid. Conjugates in the alumina-treated CSF were hydrolyzed by heating in HClO$_4$ at pH 1. The released catechols were adsorbed onto alumina and eluted with acetic acid. Aliquots of the alumina eluates were analyzed for dopamine by HPLC as described by Muldoon et al. (1979). The chromatographic system included a μBondapak C$_{18}$ column and a modified electrochemical detector (Moyer et al., 1979). The mobile phase was 0.07 M NaH$_2$PO$_4$ at pH 4.8 containing 0.2 mM Na$_2$EDTA, heptanesulfonate (0.5, 1.0, or 2.0 mM), and methanol (6% or 8%). Recovery of dopamine (500 pg) added to artificial CSF and carried through the entire procedure including the alumina extractions was 75%. Data reported have been corrected for this recovery. Recovery of released dopamine was 71% from dopamine-3-0-sulfate and 79% from dopamine-4-0-sulfate when heating in acid was used for hydrolysis.

RESULTS

Comparison of Methods

Both the radioenzymatic assay and HPLC were specific for dopamine as indicated by the absence of interference from other catechols which might be present in human ventricular fluid. Both free dopamine and free norepinephrine could be measured with the radioenzymatic assay. The HPLC assay was not suitable for measurement of norepinephrine in extracts of CSF, however, because the peak for norepinephrine did not separate adequately from the solvent injection peak (Figure 1).

The sensitivities of the methods were comparable (amounts giving values twice the blank were about 4 pg for norepinephrine and 40 pg for dopamine in the radioenzymatic assay and 50 pg of dopamine/ml of CSF by HPLC). Because constituents in CSF inhibited COMT, the sensitivity of the radioenzymatic assay was not improved by concentrating the CSF into a smaller volume. For HPLC, we extracted a 2 ml aliquot of CSF with alumina and injected a quarter of the alumina eluate onto the column; the sensitivity could be improved readily either by extracting a larger volume of CSF with alumina or by injecting a greater proportion of the sample onto the column. The alumina adsorption prior to HPLC had the further advantage of permitting separate determinations of free and conjugated dopamine to be carried out in the same aliquot of CSF. Alumina, in the amount we used,

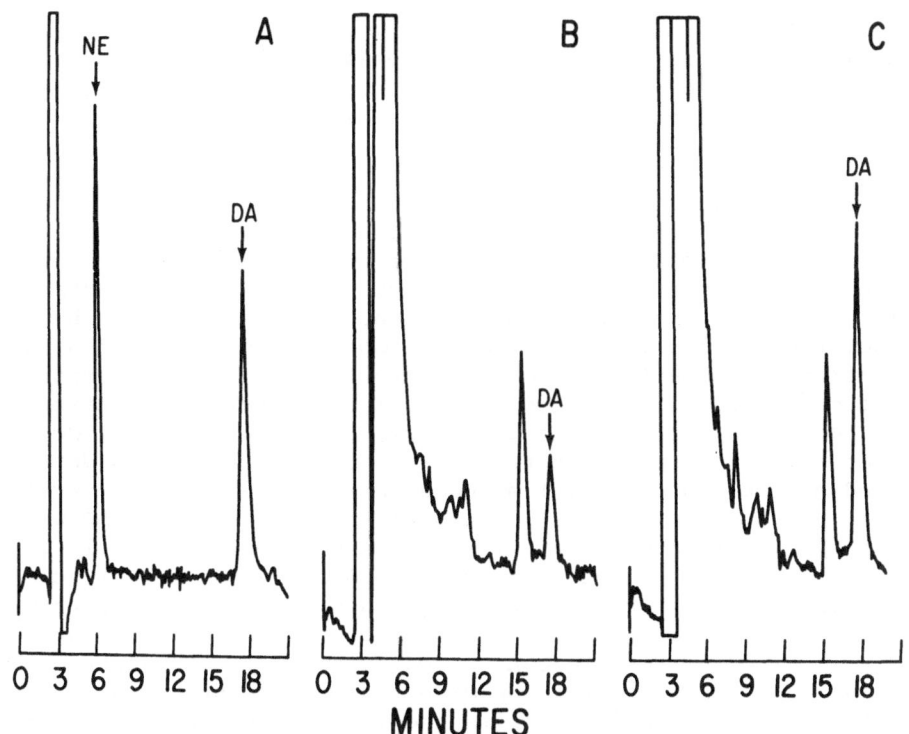

Figure 1. Chromatograms of alumina extracts of human ventricular
fluid after removal of free catechols and then hydrolysis in acid

    A. Norepinephrine (NE) and dopamine (DA): 250 pg
        in boric acid injected on the column directly
    B. Hydrolyzed ventricular fluid from patient no. 4
        who was never treated with L-DOPA
    C. Hydrolyzed ventricular fluid from patient no. 2
        who had Parkinson's disease treated with L-DOPA
        until 47 hours before surgery

Column: µBondapak C_{18}; 4 mm x 30 cm
Solvent: 0.07 M NaH_2PO_4 at pH 4.8 containing 0.2 mM Na_2EDTA,
        1.0 mM heptanesulfonate, and 6% methanol

inhibited COMT activity thus precluding use of alumina eluates
in our radioenzymatic assay.

    Although recovery of dopamine was quantitative, hydrolysis
by lyophilization in $HClO_4$ destroyed norepinephrine, making this
method of hydrolysis unsuitable for determination of conjugates

of norepinephrine in CSF. Hydrolysis by heating in acid was less
destructive to norepinephrine than lyophilization but release of
dopamine from dopamine sulfate was less when the samples were
heated instead of lyophilized. The presence of large amounts of
buffer salts prevented use of lyophilization for hydrolysis of
conjugates in alumina treated CSF.

Ventricular Fluid Studies

With the exception of one patient (no. 8), who had recently
sustained a severe head injury, free dopamine was barely detectable in unhydrolyzed ventricular fluid (Table 1). In contrast,
all of the ventricular fluids contained a compound with the
chromatographic characteristics of dopamine after they had been
hydrolyzed by heating in acid (Figure 1). The concentration of
this compound was increased in ventricular fluid from patients
who had taken the dopamine precursor, L-DOPA. The total dopamine
(free + conjugated) concentrations in ventricular fluid determined by the two different methods showed an excellent correlation ($r = 0.94$; $p < 0.001$), although the values with HPLC were
slightly lower, consistent with the less efficient hydrolysis
when the CSF was heated in acid instead of lyophilized (Table 1).
A patient with an epidural hematoma caused by a recent cerebral
trauma had increased levels of free norepinephrine (116 pg/ml)
as well as both free and conjugated dopamine in his ventricular
fluid. The ratio of free dopamine to norepinephrine was 4.5
in this patient consistent with the lesser norepinephrine content
of structures lining the lateral ventricles.

TABLE 1. Concentrations (pg/ml) of free and conjugated dopamine
in human ventricular fluid obtained with two different methods

Patient diagnonis	Radioenzyme Free	Total	HPLC Free	Conj	Total
1. Parkinson-off L-DOPA 23 Hr	---	2784	Trace	2003	2003
2. Parkinson-off L-DOPA 47 Hr	---	1427	---	1076	1076
3. Parkinson-off L-DOPA 65 Hr	---	518	70	279	349
4. Post-trauma encephalopathy	43	428	50	290	340
5. Cerebral palsy	---	223	70	254	324
6. Cerebral palsy	---	428	Trace	139	139
7. Multiple sclerosis	10	275	Trace	244	244
8. Acute cerebral trauma	516	880	641	577	1219

Data from Sharpless et al. (1980b)

DISCUSSION

The close agreement between the values obtained by two totally different methods strongly suggests that the compound we have measured in hydrolyzed ventricular fluid of man is, in fact, dopamine. To our knowledge, conjugated dopamine has not been previously reported in CSF from patients who have never taken L-DOPA. In earlier work, we found a substantial amount of a conjugate of dopamine, probably dopamine sulfate, in lumbar CSF of two patients with Parkinson's disease who were taking Sinemet, a 10:1 mixture of L-DOPA and a peripheral decarboxylase inhibitor (Tyce et al., 1980). Dopamine sulfate is one of the most abundant metabolites of L-DOPA present in plasma of patients with Parkinson's disease during treatment with L-DOPA (Tyce et al., 1974). Studies in dogs suggested a peripheral source for the dopamine conjugate present in CSF during L-DOPA administration (Tyce et al., 1980). The finding that a conjugate of dopamine occurs in CSF of patients never treated with L-DOPA lends support to the idea that the conjugate in CSF had come from brain and not from peripheral tissues. Rapid conjugation of released dopamine may explain why free dopamine has been previously unmeasureable in CSF of man.

Conjugation with sulfuric acid has long been recognized as a major metabolic pathway of the catecholamines and their metabolites in man and in laboratory animals (Williams, 1959; Goodall and Alton, 1968). Because the conjugates were difficult to measure and initially assumed to be unimportant degradation products, the presence and possible significance of the catecholamine conjugates in biological fluids other than urine have only recently received attention. In man, over 80% of circulating norepinephrine and essentially all of the plasma dopamine are conjugated (Imai et al., 1970; Buu and Kuchel, 1977; Johnson et al., 1980). Conjugates of 3-methoxy-4-hydroxyphenylglycol, the principal cerebral metabolite of norepinephrine, and of dihydroxyphenylacetic acid, an acid metabolite of dopamine, have been demonstrated to occur in brain and CSF of man and laboratory animals (Gordon et al., 1976; Elchisak et al., 1977; Karoum et al., 1977). The physiological significance of these conjugates is unknown, although evidence is accumulating which suggests that sulfate conjugation may be an important inactivation pathway for several putative amine neurotransmitters. Phenolsulfotransferase is present in human brain (Renskers et al., 1980) and dopamine is an excellent substrate (Meek and Neff, 1973). Phenolsulfotransferase is probably located in neurons rather than in glia (Meek and Foldes, 1973) and may occur postsynaptically to neurons whose transmitters are phenols. Sulfation may compete with O-methylation for inactivation of released dopamine at dopaminergic synapses. Because the sulfate esters of dopamine are zwitterions with little net charge at physiological pH, Jenner and Rose (1973) suggested that these conjugates may easily penetrate biological membranes and thus might represent convenient transport forms destined for excretion.

HVA levels in the corpus striatum and ventricular fluid of several mamalian species are increased by a number of antipsychotic drugs (Matthysse, 1973; Sedvall, 1975; Wyatt et al., 1976). These drugs also increased release of labelled dopamine into superfusates of the caudate nucleus in vivo (Besson et al., 1973). Conjugation of dopamine might therefore be expected to assume a critically important role during chronic treatment of psychotic patients with neuroleptics. Michelot et al. (1977) reported that cultured murine neuroblastoma cells from a strain deficient in monoamine oxidase activity, when incubated with dopamine, produced about equal amounts of 3-methoxytyramine and dopamine sulfate, the respective products of the methylation and conjugation pathways. When the methylation pathway was inhibited, however, conjugation became the predominant metabolic reaction. Thus, even if conjugation were normally a minor pathway of dopamine inactivation, conjugation could assume an important role in clinical conditions or animal studies where other pathways are functioning abnormally or are pharmacologically blocked. Further studies are needed to define the role of cerebral dopamine conjugation in patients with possible disorders of dopaminergic function.

ACKNOWLEDGEMENTS

Supported by NIH grants NS 09649 and NS 09143 and by a grant from the Dystonia Medical Research Foundation.

REFERENCES

Besson, M.J., Cheramy, A., Glowinski, J., and Cauchy, C. (1973). In vivo release of ^{3}H-DA from the cat caudate nucleus. In Frontiers in Catecholamine Research, (eds. E. Usdin and S.H. Snyder), Pergamon Press, New York.

Buu, N.T. and Kuchel, O. (1977). A new method for the hydrolysis of conjugated catecholamines. J. Lab. Clin. Med., 90, 680-685.

Elchisak, M.A., Maas, J.W., and Roth, R.H. (1977). Dihydroxyphenylacetic acid conjugate: natural occurrence and demonstration of probenecid-induced accumulation in rat striatum, olfactory tubercles and frontal cortex.

Garelis, E., Young, S.N., Lal, S., and Sourkes, T.L. (1974).
    Monamine metabolites in lumbar CSF: the question of their
    origin in relation to clinical studies. Brain Res., 79,
    1-8.

Glowinski, J. and Axelrod, J. (1966). Effects of drugs on the
    disposition of H^3-norepinephrine in the rat brain.
    Pharmacol. Rev., 18, 775-785.

Goodall, McC. and Alton, H. (1968). Metabolism of 3-hydroxy-
    tyramine (dopamine) in human subjects. Biochem. Pharmacol.,
    17, 905-914.

Gordon, E.K., Markey, S.P., Sherman, R.L., and Kopin, I.J.
    (1976). Conjugated 3,4 dihydroxyphenyl acetic acid (DOPAC)
    in human and monkey cerebrospinal fluid and rat brain and
    the effects of probenecid treatment. Life Sci., 18,
    1285-1292.

Imai, K., Sugiura, M., and Tamura, Z. (1970). The presence of
    conjugated dopamine in normal human plasma. Chem. Pharm.
    Bull. (Tokyo), 18, 2134.

Jenner, W.N. and Rose, F.A. (1973). Studies on the sulphation
    of 3,4-dihydroxyphenylethylamine (dopamine) and related
    compounds by rat tissues. Biochem. J., 135, 109-114.

Johnson, G.A., Baker, C.A., and Smith, R.T. (1980). Radio-
    enzymatic assay of sulfate conjugates of catecholamines
    and DOPA in plasma. Life Sci., 26, 1591-1598.

Karoum, F., Moyer-Schwing, J., Potkin, S.G., and Wyatt, R.J.
    (1977). Presence of free, sulfate and glucuronide con-
    jugated 3-methoxy-4-hydroxyphenylglycol (MHPG) in human
    brain, cerebrospinal fluid and plasma. Brain Res., 125,
    333-339.

Levin, B.E. and Hubschmann, O.R. (1980). Dorsal column stimula-
    tion: effect on human cerebrospinal fluid and plasma
    catecholamines. Neurology, 30, 65-71.

Matthysse, S. (1973). Antipsychotic drug actions: a clue to
    the neuropathology of schizophrenia? Fed. Proc., 32,
    200-205.

Meek, J.L. and Foldes, A. (1973). Sulfate conjugates in the
    brain. In Frontiers in Catecholamine Research, (eds.
    E. Usdin and S.H. Synder), Pergamon Press, New York.

172   Meek, J.L. and Neff, N.H (1973). Biogenic amines and their metabolites as substrates for phenol sulfotransferase (EC 2.8.2.1) of brain and liver. J. Neurochem., 21, 1-9.

Michelot, R.J., Lesko, N., Stout, R.W., and Coward, J.K. (1977). Effect of S-adenosylhomocysteine and S-tubercidinylhomocysteine on catecholamine methylation in neuroblastoma cells. Mol. Pharmacol., 13, 368-373.

Moir, A.T.B., Ashcroft, G.W., Crawford, T.B.B., Eccleston, D., and Guldberg, H.C. (1970). Cerebral metabolites in cerebrospinal fluid as a biochemical approach to the brain. Brain, 93, 357-368.

Moyer, T.P., Jiang, N.-S., and Machacek, D. (1979). Analysis of urinary and plasma catecholamines by high-performance liquid chromatography with amperometric detection. In Biological/Biomedical Applications of Liquid Chromatography II, (ed. G.L. Hawk), Marcel Dekker, New York.

Muldoon, S.M., Tyce, G.M., Moyer, T.P., and Rorie, D.K. (1979). Measurement of endogenous norepinephrine overflow from canine saphenous vein. Am. J. Physiol., 236, H263-H267.

Papeschi, R., Molina-Negro, P., Sourkes, T.L., and Erba, G. (1972). The concentration of homovanillic and 5-hydroxyindoleacetic acids in ventricular and lumbar CSF. Neurology, 22, 1151-1159.

Papeschi, R., Sourkes, T.L., Poirer, L.J., and Boucher, R. (1971). On the intracerebral origin of homovanillic acid of the cerebrospinal fluid of experimental animals. Brain Res., 28, 527-533.

Peuler, J.D. and Johnson, G.A. (1977). Simultaneous single isotope radioenzymatic assay of plasma norepinephrine, epinephrine and dopamine. Life Sci., 21, 625-636.

Pletscher, A., Bartholini, G., and Tissot, R. (1967). Metabolic fate of L-(^{14}C)DOPA in cerebrospinal fluid and blood plasma of humans. Brain Res.,4, 106-109.

Portig, P.J. and Vogt, M. (1969). Release into the cerebral ventricles of substances with possible transmitter function in the caudate nucleus. J. Physiol., 204, 687-715.

Renskers, K.J., Feor, K.D., and Roth, J.A. (1980). Sulfation of dopamine and other biogenic amines by human brain phenol sulfotransferase. J. Neurochem., 34, 1362-1368.

Sedvall, G. (1975). Receptor feedback and dopamine turnover in CNS. In Handbook of Psychopharmacology 6, (eds. L.I. Iversen, S.D. Iversen, and S.H. Snyder), Plenum Press, New York.

Sharpless, N.S., Tyce, G.M., Thal, L.J., Waltz, J.M., Tabaddor, K., and Wolfson, L.I. (1980a). An acid hydrolyzable conjugate of dopamine in human ventricular fluid. Neurosci. Abstracts, 6, 442.

Sharpless, N.S., Tyce, G.M., Thal, L.J., Waltz, J.M., Tabaddor, K., and Wolfson, L.I. (1980b). Free and conjugated dopamine in human ventricular fluid. Brain Res., in press.

Tabaddor, K., Wolfson, L.I., and Sharpless, N.S. (1978a). Ventricular fluid homovanillic acid and 5-hydroxyindoleacetic acid concentrations in patients with movement disorders. Neurology, 28, 1249-1253.

Tabaddor, K., Wolfson, L.I., and Sharpless, N.S. (1978b). Diminished ventricular fluid dopamine metabolite level in adult-onset dystonia. Neurology, 28, 1254-1258.

Tyce, G.M., Sharpless, N.S., Kerr, F.W.L., and Muenter, M.D. (1980). Dopamine conjugate in cerebrospinal fluid. J. Neurochem., 34, 210-212.

Tyce, G.M., Sharpless, N.S., and Muenter, M.D. (1974). Free and conjugated dopamine in plasma during levodopa therapy. Clin. Pharmacol. Therap., 16, 782-788.

Williams, R.T. (1959). Detoxication Mechanisms: The Metabolism and Detoxication of Drugs, Toxic Substances and Other Organic Compounds, Wiley, New York.

Wolfson, L.I., Katzman, R., and Escriva, A. (1974). Clearance of amine metabolites from the cerebrospinal fluid: the brain as a "sink". Neurology, 24, 772-779.

Wood, J.H. (1980). Neurobiology of Cerebrospinal Fluid 1, Plenum Press, New York.

Wyatt, R.J., Cantor, F., Gillin, J.C., Gordon, E., Karoum, F., McCullough, D., Neff, N., Ommaya, A., Rauscher, F.P., Seaborg, J.B., and Slaby, A. (1976). Ventricular fluid metabolites of phenolic amines and catecholamines. In Trace Amines and the Brain, (eds. E. Usdin and M. Sandler), Marcel Dekker, New York.

174   Ziegler, M.G., Lake, C.R., Wood, J.H., and Ebert, M.H. (1976).
        Circadian rhythm in cerebrospinal fluid noradrenaline of
        man and monkey. Nature, 264, 656-658.

        Ziegler, M.G., Wood, J.H., Lake, C.R., and Kopin, I.J. (1977).
        Norepinephrine and 3-methoxy-4-hydroxyphenyl glycol grad-
        ients in human cerebrospinal fluid. Am. J. Psychiat., 134,
        565-568.

Catecholamine Sulfates and Platelet Phenolsulfotransferase (PST) Activity in Human Hypertension

O Kuchel, N T Buu, P Hamet, P Larochelle and M Bourque

Clinical Research Institute, Hôtel-Dieu Hospital and University of Montreal, Montreal, Quebec, Canada

INTRODUCTION

Sulfoconjugation of catecholamines is a recently recognized yet poorly explored mechanism of inactivation of catecholamines (CA) in man. The current determinations of norepinephrine (NE) and epinephrine (E) measure only the free fraction of plasma NE and E (approximately 20% of the total) while the conjugated fractions accounting for the remaining approximately 80% of circulating NE and E can be determined only after hydrolysis by which free NE and E are liberated from the conjugated form (Buu & Kuchel, 1977). The process of conjugation is dependent on the type of catecholamine, its affinity to, activity and degree of saturation of the sulfoconjugating enzyme phenolsulfotransferase (PST) and its accessibility to the sulfoconjugating sites; sulfate concentration does not appear to represent a limiting factor of sulfoconjugation (Wettering et al., 1979). Our previous studies suggested that sulfoconjugation occurs during passage of catecholamines through the blood in man (Kuchel et al., 1980) and dog (Unger et al., 1980). The platelets appear to be an important source of PST (Hart et al., 1979) responsible for the sulfoconjugation that takes place in blood.

THE NORMAL SULFOCONJUGATING PROCESS AND ITS SITES

Radioenzymatic determinations of conjugated NE and E calculated from estimations of free NE and E liberated after hydrolysis have shown (Table 1) that for normal subjects the values obtained following three methods of acid hydrolysis - lyophilization in acids (Buu & Kuchel, 1977) acid hydrolysis (Da Prada, 1980 and Nagel & Schümann, 1980) are similar to those following an enzymatic hydrolysis by sulfatase (Johnson et al., 1980). Although the absolute values of free and conjugated plasma and uri-

TABLE 1

Human plasma free and conjugated NE and E (mean ± SE) radioenzymatically determined after hydrolysis of conjugated
catecholamines by four methods

Hydrolysis method (mean ± SE)	Free NE	Conj. NE	% conj.	free E	conj. E	% conj.
Lyophilization in acids (Buu & Kuchel, 1977)	0.23 ±0.05	0.57 ±0.2	71	0.04 ±0.02	0.32 ±0.1	88
Acid Hydrolysis (Da Prada, 1980)	0.5 ±0.1	1.3 ±0.4	71	0.05 ±0.01	0.30 ±0.1	89
Acid Hydrolysis (Nagel & Schümann, 1980)	0.33 ±0.04	1.0 ±0.2	68	0.08 ±0.07	0.45 ±0.1	80
Sulfatase Hydrolysis* (Johnson et al. 1980)	0.67	1.29	66	0.08	0.44	83

* means of 3 human plasma pools after subtracting the free from total NE and E values
respectively.

nary NE and E may vary with sympathetic activity (e.g. due to the
stress of sampling (Lake et al., 1976)), the ratio of NE:E of the
degree of conjugation remains similar for all 3 methods, plasma
E being more highly conjugated (80-89%) than NE (66-71%). The
higher degree of E conjugation probably reflects a higher dependence of E on sulfoconjugation. As a circulating hormone in the
traditional sense (Cryer, 1980), E is more accessible to sulfoconjugating sites than NE, which is a typical neurotransmitter.
It is however remarkable that free NE with an only limited
"spillover" into the circulation has normally higher plasma concentrations than free E. This is probably due to an at least 4
times higher release of NE than E from endogenous sources (Kopin
1979).

The tissue distribution of PST activity determined by DA sulfation (Table 2) shows that in the dog, in which dopamine is sulfoconjugated as in the man (Merits, et al., 1973), some organs
which are rich in sympathetic nervous terminals (such as the
heart and spleen) have low PST activity. This does not support
the hypothesis (Meek, Foldes, 1974) that PST is present in highly innervated organs at the postsynaptic membrane. Despite high
PST activity in vitro in the liver, the absence of arteriovenous
differences for free and conjugated DA across the liver in dog
(Unger et al., 1980) suggests that the liver, well known to sulfoconjugated exogenous phenolic substances, does not contribute
to sulfoconjugation of circulating endogenous phenols; sulfoconjugation probably occurs during their passage in the blood by

TABLE 2   177

Tissue PST activity (units per mg protein) mean ± SE

(1 unit = the PST activity catalysing the sulfation of one pmole

of substrate per minute)

	Dog	Man
Liver	495 ± 109	-
Kidney	132 ± 41	1.8 ± 0.9 (SD)*
Lung	60 ± 26	-
Jejunum	41 ± 6	81 ± 38 (SD)*
Heart	10 ± 2	-
Spleen	N.D.	-
Pancreas	N.D.	-
Adrenals	17 ± 18	61
Platelets	70	55 ± 27, 83 ± 29 (SD)**
Brain	6	-

N.D. Non detectable
*Anderson, Weinshilboum 1980a, ** Anderson, Weinshilboum, 1980.

platelets capable to accumulate free DA (Bouillon and O'Brien,
1970). Affinity studies of human platelet PST have shown (Figure
1) that the highest affinity for platelet PST is in the order DA

Figure 1. Human platelet PST affinity towards epinephrine, norepinephrine and dopamine; protein concentrations, ● 50 μg/ml, o
65 μg/ml

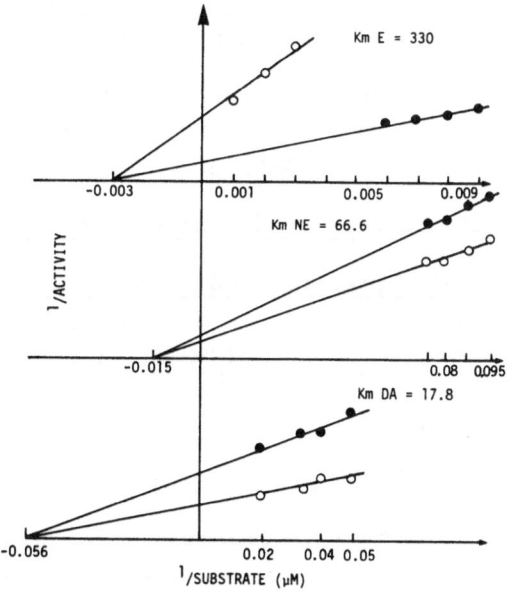

(highest), NE, E (Abenhaim et al., 1981). The fact that despite its 5 times lower affinity for PST, E is more highly conjugated than NE indicates the overriding importance of factors other than PST activity in determining the degree of sulfoconjugation.

CATECHOLAMINE SULFATES AND PLATELET PST IN HUMAN HYPERTENSION

Although our knowledge of sulfoconjugation in man is limited it was of interest to evaluate the sulfoconjugation of CA in a condition in which the role of CA is most controversial, i.e. human essential hypertension. By analogy with the tyramine sensitive migraine in which a defect of sulfoconjugation has been postulated (Youdim et al., 1971) we found in a form of essential hypertension mimicking pheochromocytoma (increased sympathetic tone) very low or even undetectable concentrations of conjugated CA (the sum of NE and E) contrasting with an increased plasma free CA baseline concentration which was hyperresponsive both to stress and to glucagon (Kuchel et al., 1981). This suggested that a defficient sulfoconjugation system left more NE and E in the free state so that the adrenergic receptors were more exposed to the action of CA, and as a result, induced the symptoms and signs of sympathetic hyperactivity.

In view of the differences between the degree of NE and E conjugation and the fact that only the sum of NE and E was determined in the previous study, we assayed plasma free NE and E separately both with regard to baseline values and responses to stimuli such as glucagon. We grouped 38 essential hypertensive patients,who had NE and E separately determined, according to the principle previously described (Kuchel et al., 1981), considering those with conjugated E lower than 0.09 ng/ml and conjugated NE lower than 0.10 ng/ml (2 SD below the mean for control subjects) to have low conjugated E, NE respectively. We found (Table 3) that patients with low conjugated E had higher free plasma E than patients with normal conjugated E whereas free

TABLE 3

Separate effects of the degree of conjugation of E and NE on free

NE and E, pulse pressure and pulse rate (mean ± SE)

	Plasma E ng/ml		Plasma NE ng/ml		Pulse pressure	Maximum Pulse rate beats/min.
	conjugated	free	conjugated	free		
Normal conjugated E (n = 16)	0.49 ± 0.21	0.07 ± 0.02	0.30 ± 0.07	0.20 ± 0.03	57 ± 5.4	85 ± 2
Low conjugated E (n = 22)	0.01 ± 0.004	0.14 ± 0.02	0.18 ± 0.05	0.28 ± 0.04	71 ± 7.3	101 ± 6
Normal conjugated NE (n = 22)	0.32 ± 0.16	0.11 ± 0.02	0.39 ± 0.06	0.24 ± 0.04	63 ± 7	92 ± 3
Low conjugated NE (n = 16)	0.07 ± 0.03	0.10 ± 0.03	0.02 ± 0.01	0.26 ± 0.05	67 ± 7	96 ± 8

*p < 0.05 or less

NE was not significantly different between patients with normal
and those with low conjugated E or NE. As far as the relationship
between the conjugation status of E and NE on one hand and hemodynamic parameters on the other is concerned, patients with low
conjugated E had a higher maximum pulse rate than patients with
normal conjugated E.

When all patients were combined without grouping, there was
an overall negative correlation between free E and the degree of
E conjugation (r = -0.62, p < 0.001) and to a lesser degree between free NE and the degree of NE conjugation (r = -0.36, p <
0.01). There was an overall positive correlation (Table 4) be-

TABLE 4

Correlation coefficients between free E and NE as well as

their degree of conjugation and some hemodynamic indices

| | PULSE RATE | | PULSE PRESSURE | |
	baseline	maximal	(systolic – diastolic difference)	diastolic blood pressure
Plasma free E	r = 0.44 p < 0.003	r = 0.63 p < 0.001	r = 0.28 p < 0.04	r = 0.26 N.S.
% of conjugated E	r = -0.36 p < 0.01	r = -0.79 p < 0.001	r = -0.12 N.S.	r = -0.12 N.S.
Plasma free NE	r = 0.25 N.S.	r = 0.26 N.S.	r = 0.03 N.S.	r = 0.29 p < 0.04
% of conjugated NE	r = -0.16 N.S.	r = -0.15 N.S.	r = -0.01 N.S.	r = -0.35 p < 0.01

N.S.: non significant.

tween plasma free E and the baseline pulse rate and pulse pressure as well as the maximum pulse rate. The degree of E conjugation was negatively correlated with the baseline and maximum
pulse rates. None of these correlations was observed with free
NE, but the latter correlated positively with the baseline diastolic BP to which the degree of conjugated NE was negatively related.

These results indicate that the effect of the previously described CA sulfation defect (Kuchel et al., 1981) on free NE and
E is predominantly on E. Higher free E levels due to its inade-
quate conjugation may result the exposure of adrenergic recep-

tors to higher free E concentration. The main elements of this "β-hyperadrenergic hypertension" - increased heart rate and pulse pressure positively related to plasma free E, are compatible with this action of free E. This explanation is also supported by the finding that most of the patients with low conjugated E had free E exceeding the threshold concentrations resulting in an increase in pulse rate (0.075 ng/ml) and systolic blood pressure (0.1 ng/ml) some of them also the concentration resulting in a decrease of diastolic blood pressure (0.2 ng/ml) (Clutter et al., 1980). The latter two actions resulted in an increased pulse pressure (systolic-diastolic difference). Almost all patients with normal conjugated E had free E values below these thresholds. In contrast to free E there was a weak positive correlation between free NE and diastolic BP; free plasma NE levels, whether conjugated NE was low or normal, were however far below the hemodynamic threshold of NE of 1.5 ng/ml (Silverberg et al., 1978) and are thus probably unrelated to hypertension but reflect merely a "spillover" of the neurotransmitter into the circulation. Despite these differences between free E and NE there appears to be a common factor determining the degree of conjugation of both CA as suggested by positive correlation between the levels of NE and E that are conjugated ($r = 0.42$, $p < 0.005$).

As to this common factor resulting in "unconjugated (free) hyperepinephrinemia" the finding that platelet PST activity reflects the PST activity of some organs such as the gut and kidney (Anderson et al., 1980a) supports the use of platelets as an easily accessible marker of PST activity. Measurements of platelet PST activity by a radioenzymatic method using dopamine as a substrate (Abenhaim et al., 1981) initially suggested very low or absent PST activity in some patients. When repeating the measurements by the method using 3-methoxy- 4-hydroxyphenylglycol (MHPG) as substrate (Anderson & Weinshilboum, 1980) in other patients belonging to the two subgroups we again found some low values but no significant differences between the groups (Table 5). The PST activity determined with DA substrate and MHPG substrate correlated positively ($r = 0.73$, $p < 0.001$).

When trying to correlate the individual PST activities and the absolute concentration of conjugated DA, NE and E we were unable to find any relationship between platelet PST activity and plasma conjugated DA and E ($r = 0.16$ and -0.11). There was however a positive correlation between platelet PST activity and conjugated NE ($r = 0.40$, $p < 0.006$). Until we accumulate more data it is difficult to decide whether NE has a special relationship to platelet PST, the latter possibly determining the sulfoconjugating shunt of NE and so indirectly the concentrations of free NE. In any case the complexities of the sulfoconjugation make it improbable that PST activity should be the only determinant of the level of any conjugated CA. This does not

TABLE 5 181

Platelet PST activity in control and hypertensive subjects

(mean ± SD)

	Control Subjects	ESSENTIAL HYPERTENSION WITH CONJUGATED NE + E	
		Low (< 0.23 ng/ml)	Normal (> 0.23 ng/ml)
Platelet PST activity:			
with DA substrate*	55 ± 27	37 ± 8	58 ± 25
pmol /min/mg protein (range)	(19 - 90)	(0 - 70)	(12 - 129)
with MHPG substrate **	40 ± 10 83 ± 29**	18 ± 8	32 ± 8
pmol /min/mg protein (range)	(2.2 - 89) (34 - 193)	(2 - 30.4)	(1.6 - 87)

* Abenhaim et al., 1981
** Anderson, Weinshilboum 1980.

exclude however the possibility that marginally decreased PST activity or its saturation by another CA may result in a conjugation defect of a CA highly dependent on sulfoconjugation and having a low affinity to PST such as epinephrine. Plasma dopamine sulfates have been found to be elevated in essential hypertension and this even in patients who have low or absent conjugated NE and E (Kuchel et al., 1981). This may be due to a combination of high affinity to platelet PST and easy accessibility of DA to sulfoconjugating sites. Only further determinations of platelet PST activity and of other components of the sulfoconjugating and deconjugating system will help us to understand whether a defect of PST which in some of our patients seems likely to be genetically determined, may be responsible for this form of labile hypertension often imitating pheochromocytoma.

If an alteration in the process of sulfoconjugation is able to modify the clinical presentation of essential hypertension to imitate pheochromocytoma, the opposite would be true for some true pheochromocytoma patients whose clinical presentation might be modified by high sulfoconjugation to imitate essential hypertension. We were able to demonstrate (Kuchel et al., 1980) that a subgroup of patients with very high conjugated NE + E had less free NE and E overflow during hypertensive crises and less urinary excretion of VMA and free NE + E than patients with normal conjugated NE + E. Our data on platelet PST and tumor PST activity in these patients are too limited at the moment to decide whether or not PST activity was induced by the hypersecretion by these patients either in the tumor (as suggested by the release of conjugated CA from the tumor itself (Kuchel et al., 1980)) or in the circulation.

182   Another form of hypertension in which the degree of sulfoconjugation of catecholamines may be relevant to the pathophysiology
is <u>primary aldosteronism</u>. We have found that spontaneous, nyct-
hemeral or dexamethasone-induced variations in plasma aldosterone
were indirectly correlated with changes in free or conjugated
plasma DA in the opposite direction (Kuchel et al, 1980a) which
suggested an inhibitory action of DA on aldosterone secretion.
After dexamethasone administration we found an increase in plasma free DA ($p < 0.004$) and free NE ($p < 0.01$) while conjugated
DA and NE decreased in most cases; this resulted in a decrease
in the degree of plasma DA conjugation (from 94 to 81%, $p < 0.05$)
and NE conjugation (from 89 to 61%, $p < 0.01$). It is not clear
whether this is an effect of dexamethasone, or of the ACTH suppression induced by it, on the sulfoconjugation and how it rela-
tes to the dexamethasone-induced increase in some tissue PST activity (Maus et al, 1980). In any case it demonstrates that
some factors other than those previously enumerated can affect
the process of sulfoconjugation and so indirectly the biological
activity of the relevant CA. An example of an action which may
be determined by the degree of conjugation of DA is the dexamethasone-induced suppression of aldosterone in primary aldostero-
nism (Carey et al., 1980) which may be mediated, at least in
part, by increasing free plasma DA. Sulfoconjugation of phenolic substances in general may thus not only be a determinant of
the biodegradation of many drugs having a phenolic structure
(Davies, 1975) but also a site of action of some homeostatic regulatory factors determining the ratio of biologically active
free and inactive conjugated catecholamines.

    Hypertension appears to be a fruitful area of exploration of
the endogenous bio-disponitiblity of CA as a function of their
sulfoconjugation, one of its determinants, among others. Many
remains to be learned however about the genetic and environmental modulation of the sulfoconjugation and its effect on the biological action of CA.

    This work was supported by a grant from the Medical Research
Council of Canada to a multidisciplinary research group in
hypertension.

REFERENCES

Abenhaim L, Romain Y, Kuchel O. (1981). Platelets Phenolsul-fotransferase and Catecholamines: Physiological and Pathological Variations in Humans. Can. J. Physiol. Pharmacol. 59:000-000.

Anderson R.J., Weinshilboum R.M. (1980). Phenolsulphotrans-ferase in Human Tissue: Radiochemical Enzymatic Assay and Biochemical Properties. Clin. Chim. Acta 103:79-90.

Anderson R., Weinshilboum R., Philips S., Broughton, D. (1980a) Human Platelet Phenolsulphotransferase (PST) Activity: Correlation with PST Activity in the Kidney and Gut. Pharmacologist 22(3): p. 301.

Bouillon D.O., O'Brien R.A. (1970). Accumulation of Dopamine by Blood Platelets from Normal Subjects and Parkinsonian Patients under Treatment with L-dopa. Br. J. Pharmacol. 39:779-788.

Buu, N.T., Kuchel O. (1977). A New Method for the Hydrolysis of Conjugated Catecholamines. J. Lab. Clin. Med. 90:680-685.

Carey R.M., Thorner M.O., Ortt, E.M. (1980). Dopaminergic Inhibition of Metoclopramide -Induced Aldosterone Secretion in Man. J. Clin. Inv. 66:10-18.

Clutter W.F., Bier D.M., Shash S.D., Cryer P.E. (1980). Epinephrine Plasma Metabolic Clearance Rates and Physiologic Threshold ofr Metabolic and Hemodynamic Actions in Man. J. Clin. Inv. 66:94-101.

Cryer P.E. (1980). Physiology and Pathophysiology of the Human Sympathoadrenal Neuroendocrine System. New Engl. J. Med. 303:436-444.

Da Prada M. (1980). Concentration, Dynamics and Fonctional Meaning of Catecholamines in Plasma and Urine. Trends in Pharmacol. Sci. 1:157-159.

Davies D.S. (1975). Pharmacokinetics of Inhaled Substances. Postgraduate Medical Journal 51(Suppl. 7):69-75.

Hart R.F., Renskers K.J., Nelson E.B, Roth, J.A. (1979). Localization and Characterization of Phenol Sulfotransferase in Human Platelets. Life Sci. 24:125-130.

Johnson G.A., Baker C.A., Smith R.T. (1980): Radioenzymatic Assay of Sulfate Conjugates of Catecholamines and Dopa in Plasma. Life Sci. 26:1591-1598.

184   Kopin I.J.: (1979). Biochemical Assessment of Peripheral
Adrenergic Activity, In The Release of Catecholamines from Adrenergic Neurons (David M. Paton eds.) Pergamon Press, Oxford, pp
355-371.

        Kuchel O., Buu N.T., Fontaine A., Hamet P., Beroniade V.,
Larochelle P. and Genest J. (1980). Free and Conjugated Plasma
Catecholamines in Hypertensive Patients with and without Pheo-
chromocytoma. Hypertension 2:177-186.

        Kuchel, O., Buu N.T., Vecsei P., Bourque M., Hamet P., and
Genest J. (1980a): Are Plasma Aldosterone Surges in Primary Al-
dosteronism Due to a Loss of an Inhibitory Dopaminergic Control?
J. Clin. Endocrinol. Metab. 51(2)337-344.

        Kuchel O., Buu N.T., Hamet P., Larochelle P., Bourque, M.
and Genest J. (1981). Essential Hypertension with Low Conjugated
Catecholamines Imitates Pheochromocytoma. Hypertension 3:000-
000.

        Lake C.R., Ziegler M.G., Kopin I.J. (1976). Use of Plasma
Norepinephrine for Evaluation of Sympathetic Neuronal Function
in Man. Life Sci. 18:1315-1326,

        Maus T.P., Anderson R.J., Weinshilboum R.M. (1980). Effect
of Dexamethasone on Rat Phenolsulphotransferase (PST) Activity.
Pharmacologist 22(3): p. 301.

        Meek J.L., Foldes A. (1974). Sulfate Conjugates in the Brain.
Biochem. Pharmacol. Supp. Part 1:116-118.

        Merits I., Anderson D.J., Sonars R.C. (1973). Species Dif-
ferences and Metabolism of Orally Administered Dopamine [14] C.
Drug. Metab. Disp. 1:691- 697.

        Nagel M. and Schümann H.J. (1980). A Sensitive Method for
Determination of Conjugated Catecholamines in Blood Plasma. J.
Clin. Chem. Clin. Biochem. 18:431-432.

        Silverberg, A.B., Shah S.D., Haymond M.W., Cryer P.E. (1978).
Norepinephrine Hormone and Neurotransmitter in Man. Am. J. Physiol. 234(3):E252-E256.

        Unger T., Buu N.T., Kuchel O., Schürch W. (1980). Conjugated
Dopamine: Peripheral Origin, Distribution and Response to Acute
Stres in the Dog. Can. J. Physiol. Pharmacol. 58:22-27.

        Wettering J.G., Krugsheld K.R., Mulder G.H. (1979). The
Availability of Inorganic Sulphate as a Rate Limiting Factor in
the Sulphate Conjugation of Xenobiotics in the Rat? Sulphation

and Glucuronidation of Phenol. Biochem. Pharmacol. 28:757-762.

Youdim M.B.H., Carter S.B., Sandler M., Hanington E., Wilkinson M. (1971). Conjugation Defect in Tyramine - Sensitive Migraine. Nature 230:127-128.

Clinical Studies on Platelet Phenolsulphotransferase

M Sandler, Vivette Glover, Susan M Bonham Carter, Julia Littlewood
and Glen Rein

Bernhard Baron Memorial Research Laboratories and Institute of Obstetrics and
Gynaecology, Queen Charlotte's Maternity Hospital, Goldhawk Road,
London W6 0XG, UK

As pointed out in the Introduction (Sandler, this volume), the identification of phenolsulphotransferase (PST) in the human platelet has given us, potentially at least, another reflection of events in the rest of the body, possibly including the brain. And with the identification of multiple forms of this enzyme, as described by Rein et al. (this volume), the possibility of obtaining useful clinical information has been greatly increased.

Rein et al. (1981a) observed that dichloronitrophenol is a selective inhibitor of phenol sulphation in platelets. This finding suggests that phenol, on the one hand, and the monoamines and their degradation products, on the other, are metabolized in the human platelet by two separate active sites, which we have tentatively called the "P" (for phenol) site and the "M" (for monoamine and monoamine metabolites) site. In order to determine whether these sites are controlled together or independently, we (Bonham Carter et al., 1981a) have measured platelet PST activity towards phenol, dopamine, tyramine and 4-hydroxy-3-methoxyphenylglycol (HMPG) in 30 subjects. For this and all the studies described in this chapter a radioactive assay was employed based on that of Foldes and Meek (1973). There was a high degree of correlation between dopamine and tyramine (correlation coefficient 0.78), between HMPG and dopamine (correlation coefficient 0.82) and between HMPG and tyramine (correlation coefficient 0.88) ($p \ll 0.001$ in all; Figure 1).

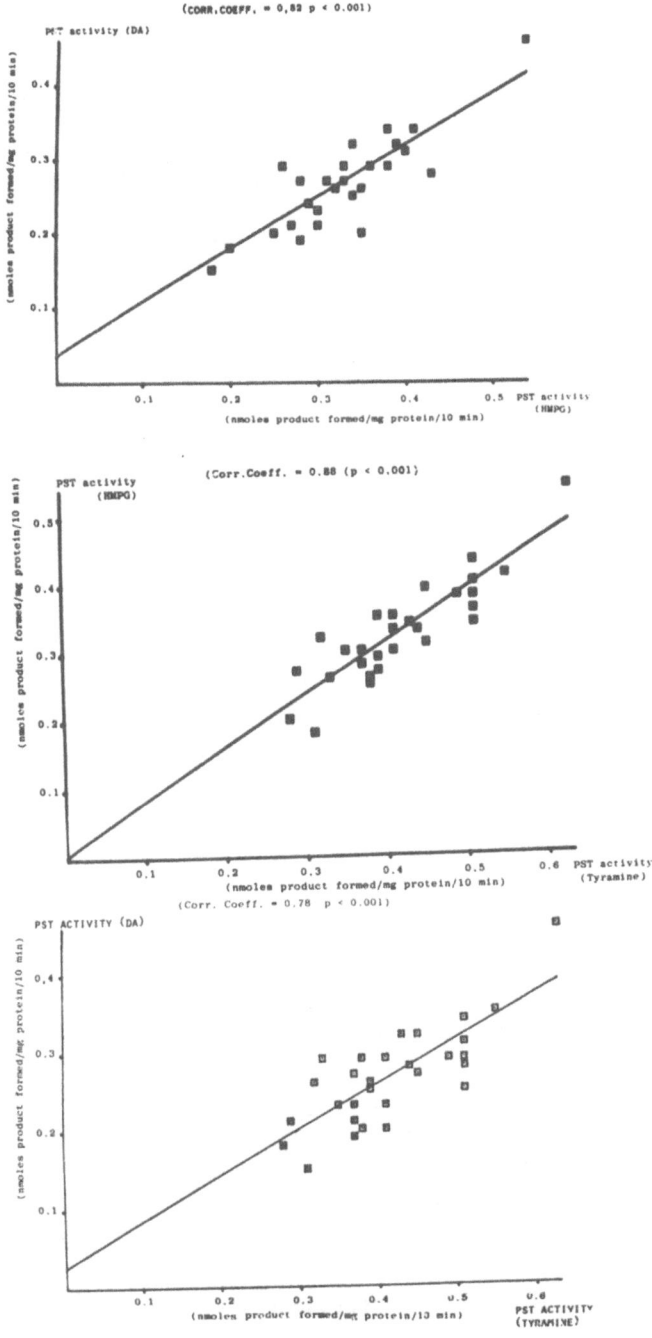

Figure 1

Significant correlations between tyramine, dopamine and HMPG conjugated by human platelet PST (Bonham Carter et al.,1981a).

188

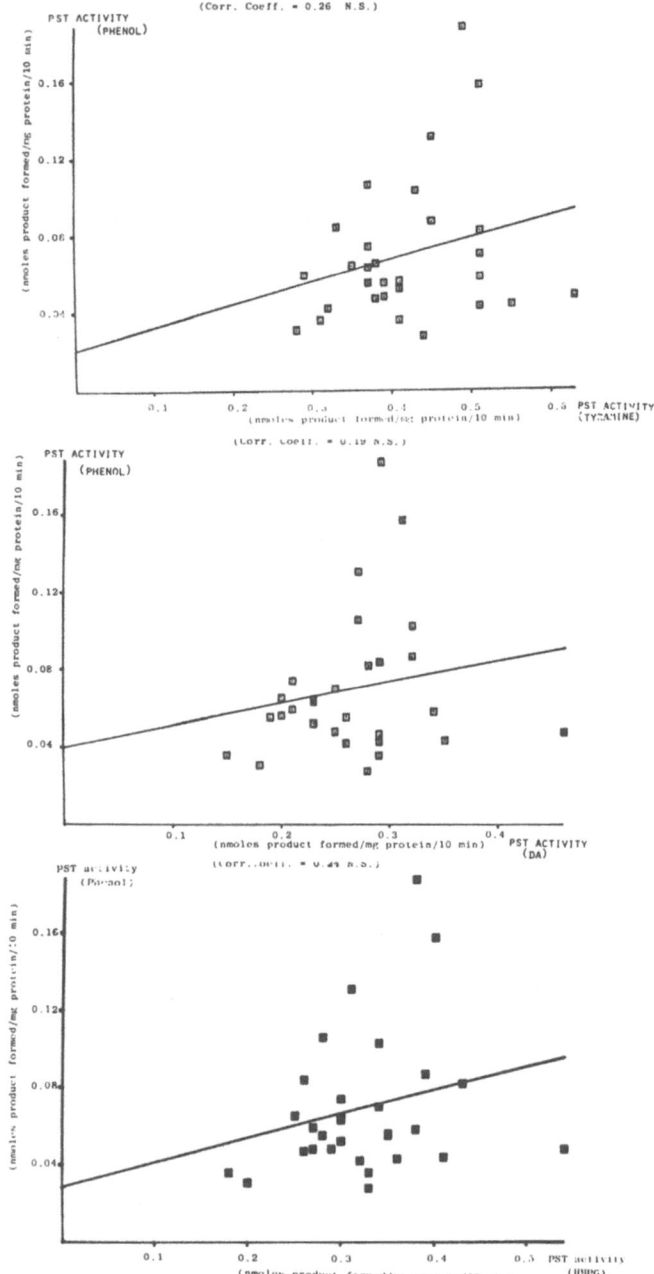

Figure 2.

Lack of correlation between phenol and tyramine, dopamine and
HMPG conjugation by human platelet PST (Bonham Carter et al.,
1981a).

Similar results have been obtained by Anderson et al.
(1981). However, when dopamine and phenol, tyramine and
phenol and HMPG and phenol, metabolizing activities
were compared, there was no significant correlation
(correlation coefficients 0.19, 0.26 and 0.24 respectively; Figure 2).

These data provide further evidence of the presence
of at least two distinct PST enzymes in the human platelet, and show clearly that they possess different con-
trolling mechanisms. One other point of interest emerg-
es: the activity range differs with the different substrate groups. When dopamine (at 30 μM), tyramine (at
133 μM) and HMPG (at 800 μM) were employed, there was
only a 2 to 3-fold variation (0.15 - 0.46, 0.28 - 0.63
and 0.18 - 0.54 nmoles product formed /mg protein/ 10
min respectively). When phenol (at 30 μM) was used as
substrate, there was an approximately 6-fold variation
of platelet activity in the same population (0.03 -
0.19 nmoles product formed/mg protein/10 min).

In this group of 30 subjects, no significant difference in either M or P activity could be detected between males and females: the observed means ($\pm$ SEM) for
the three M substrates employed, dopamine, tyramine and
HMPG in males and females were, respectively, $0.27 \pm
0.01$ and 0.26 ± 0.02; 0.43 ± 0.02 and 0.41 ± 0.02; and
0.33 ± 0.01 and 0.32 ± 0.02. P activity was 0.08 ± 0.01
and 0.06 ± 0.01. The ages of the subjects ranged from 20
to 70 yrs but there was no correlation of age with activity in respect of any of the substrates employed.

Because our discovery of the P enzyme is very recent,
we have had little opportunity of investigating its separate activity in human disease states and most of the
data presented below, therefore, apply to activity of
the M enzyme.

KINETIC STUDIES OF THE M ENZYME

Human platelet PST was assayed with different concentrations of tyramine and of PAPS (Bonham Carter et
al.,1981b). The results are shown in Figures 3 and 4.

Apparent K_m values have been calculated by the direct linear plot (Eisenthal and Cornish Bowden,1974).
That for tyramine was 17 μM (mean of two experiments)
and for PAPS, 0.14 μM (mean of two experiments). The
apparent K_m value for tyramine obtained by this method
is somewhat lower than that obtained by other workers

190

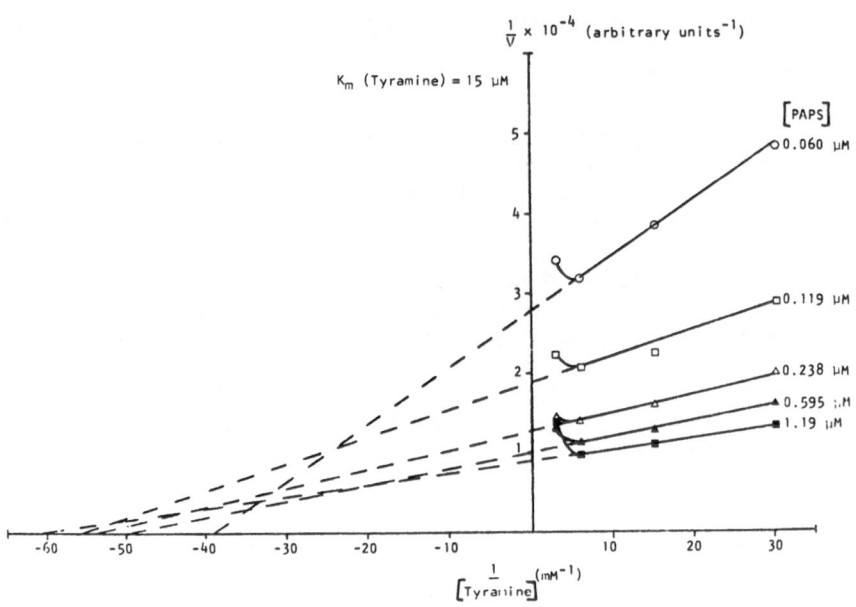

Figure 3
Double reciprocal plots of human platelet PST activity versus tyramine concentration at constant concentrations of PAPS. A second
experiment gave similar results (Bonham Carter et al.,1981b).
Lines were reconstructed by the direct linear plot method (Eisenthal and Cornish Bowden,1974).

using Lineweaver-Burk plots (Rein et al.,1981b; Hart
et al.,1979). The apparent K_m for PAPS agreed very closely with that obtained by Anderson and Weinshilboum
(1980) also using the direct linear plot.

STABILITY

For measurements to be clinically useful, it seemed
necessary to determine the stability of the platelet
enzyme, for it is not always possible to process all the
samples in a comparative series simultaneously, however
desirable this might be.

It became obvious that the stability of platelet PST
depends, to a large extent, on the medium in which it
is stored (Bonham Carter et al.,1981b). In Fig.5, the
activity with time of aliquots of washed pooled platelet samples stored deep-frozen (-20°C) in phosphate buffer (10 mM, pH 7.4), isotonic sucrose or isotonic saline

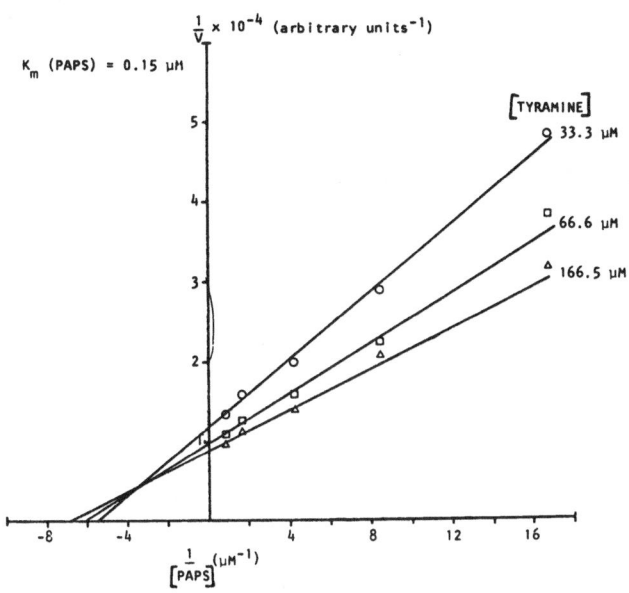

Figure 4
Double reciprocal plots of human platelet PST activity versus
PAPS concentration at constant concentrations of tyramine. A sec-
ond experiment gave similar results (Bonham Carter et al.,1981b).
Lines were reconstructed by the direct linear plot method (Eisen-
thal and Cornish-Bowden,1974).

is demonstrated. Different aliquots were assayed over a
period of 60 days. Whereas the enzyme appeared to be
stable in samples stored in buffer or sucrose, those
stored in saline lost a substantial proportion of their
activity which the addition of dithiothreitol (8 mM) to
the assay mixture failed to restore completely.

Thermostability studies of the human platelet enzyme
are shown in Fig.6. When preparations were exposed to
a variety of different temperatures for 20 min periods
prior to assay it became apparent that the M enzyme
loses most of its activity at $40^{\circ}C$; even at $35^{\circ}C$, substantial activity was lost (Bonham Carter et al.,1981b).
Once again, such activity could not be restored by
dithiothreitol (8 mM). It is obvious that great care
must be taken with temperature control during the preparation of platelets for PST studies.

192

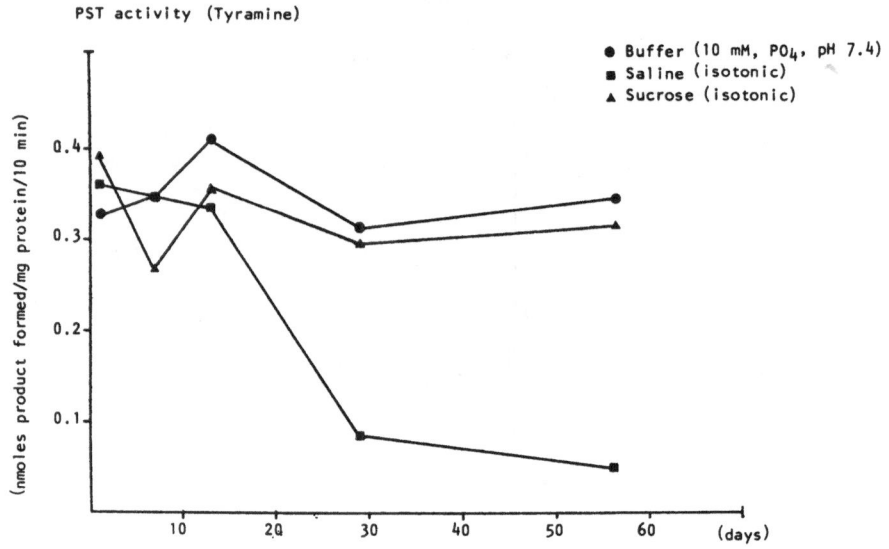

PST activity (Tyramine)

● Buffer (10 mM, PO4, pH 7.4)
■ Saline (isotonic)
▲ Sucrose (isotonic)

(nmoles product formed/mg protein/10 min)

0.4

0.3

0.2

0.1

10   20   30   40   50   60   (days)

Figure 5
Stability of human platelet PST activity with storage time in iso-
tonic saline, isotonic sucrose and phosphate buffer (10 mM, pH 7.4)
Bonham Carter et al.,1981b).

TYRAMINE CONJUGATION IN DEPRESSIVE ILLNESS

Once a reliable method for the assay of platelet PST
activity became available, it seemed important to make
measurements in patients with depressive illness for
there were good grounds for thinking that there might
be an enzyme abnormality in this condition. Over a period of years, we had established firmly that a tyramine
conjugation deficit is present in depressive illness
and this defect is trait- rather than state-dependent
(Sandler et al.,1975; Bonham Carter et al.,1978). Thus,
a group of patients with severe, therapy-resistant depressive illness showed a most pronounced inability to
conjugate an oral load of tyramine and this biochemical response was unchanged after some of the patients
had undergone virtually complete remission following
leucotomy (Figure 7). To judge from the urinary output
of the major oxidatively deaminated metabolite of tyramine, p-hydroxyphenylacetic acid, there was no evidence of a compensatory increase in monoamine oxidase
being responsible and, indeed, in this particular group of patients, mean monoamine oxidase activity was not

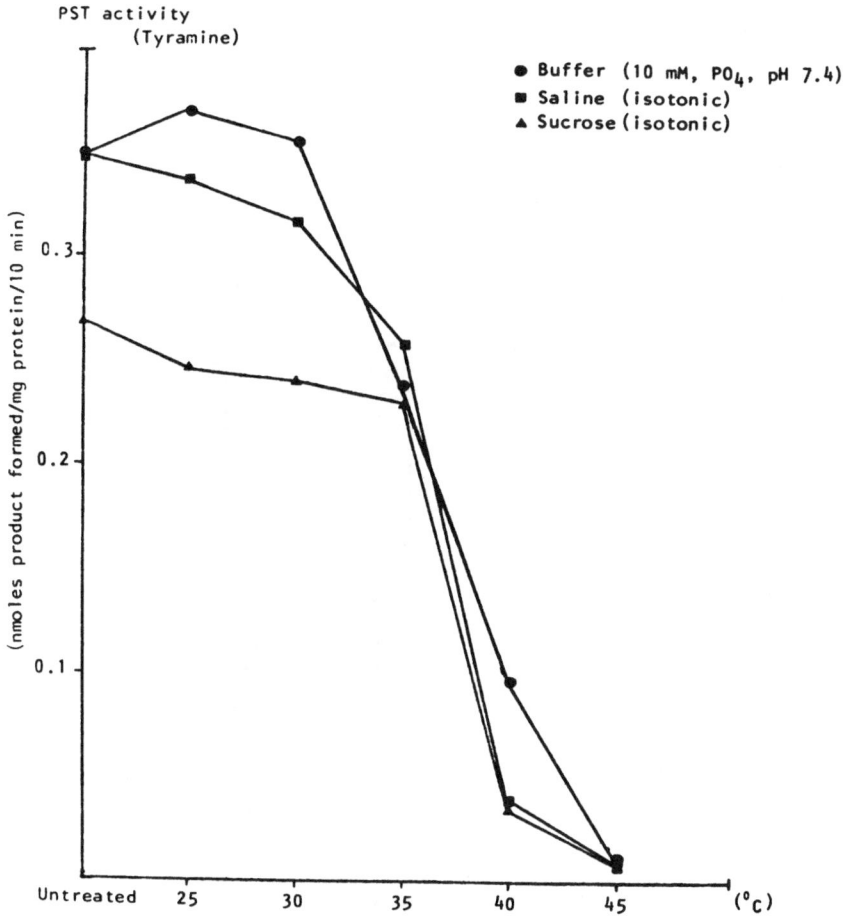

Figure 6
Thermostability of human platelet PST in isotonic saline, isotonic
sucrose and phosphate buffer (10 mM, pH 7.4) (Bonham Carter et al.,
1981b).

different from that observed in control subjects (depressed patients, 17.6 ± 1.8 nmoles product formed/mg
protein/30 min; patient controls 18.3 ± 1.9 nmoles product formed/mg protein/30 min; staff controls 16.5 ±
1.8 nmoles product formed/mg protein/30 min (Bonham Carter et al.,1981b). It should be noted, however, that
both we (Reveley et al.,1981) and others (Nies et al.,
1971,1974; Landowski et al.,1975; Mann,1979) have demonstrated a significant increase of monoamine oxidase
activity in other groups of depressive subjects so that
the matter cannot be considered completely closed.

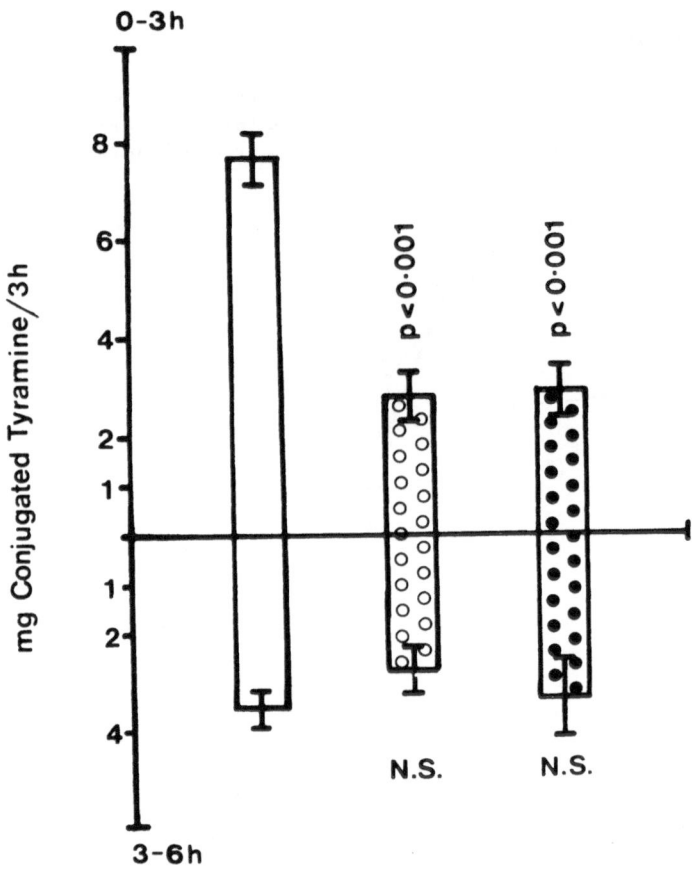

Figure 7

Excretion of conjugated tyramine (mean ± SEM) by normal subjects ☐ , severely depressed patients assessed one year after modified leucotomy as 'recovered' 🅾🅾 and severely depressed patients assessed one year after modified leucotomy as 'unchanged' ●●, after an oral tyramine load (100 mg) (Bonham Carter et al., 1978) (Reproduced by kind permission of British Journal of Psychiatry).

Although it is known that certain forms of chronic medication, classically with salicylamide, can result in depletion of body sulphate stores followed by deficient sulphate conjugation (Levy and Matsuzawa,1967), we have been able to show that the apparent conjugation deficit we have observed cannot be due merely to this cause (Bonham Carter et al.,1980a). Although sulphate loading with a large dose of cysteine substantially raised the output of sulphate-conjugated tyramine in both controls and depressed patients belonging to this therapy-resistant group, the increase in the latter (Table 1) was only

TABLE 1. Excretion of conjugated tyramine (mg/3 h mean ± s.e.mean) by control subjects and depressed patients after administration of tyramine (100 mg) and tyramine (100 mg) plus L-cysteine (9 g; 1 g/h for 9 h)

Period (h)	CONTROL SUBJECTS			DEPRESSED PATIENTS		
	Tyramine	Tyramine + cysteine	Mean % increase after cysteine	Tyramine	Tyramine + cysteine	Mean % increase after cysteine
0-3	6.46 ± 0.78	9.45 ± 0.77§§	51.4 ± 10.3	2.73 ± 0.66††	4.39 ± 0.68**†	96.0 ± 39.3
3-6	3.14 ± 0.60	4.03 ± 0.45*	39.4 ± 15.3	3.40 ± 0.97	3.66 ± 0.67	25.7 ± 15.5
Combined 0-6	9.60 ± 1.36	13.51 ± 1.06§	47.3 ± 11.9	6.13 ± 1.25	8.05 ± 1.31§§†	43.1 ± 14.4

Significant results: tyramine v tyramine + cysteine (paired t-test)

* p < 0.05
** p < 0.01
§ p < 0.005
§§ p < 0.001

Control subjects v depressed patients (Student's t-test)
† p < 0.05, †† p < 0.01

196   in a proportion similar to that in controls whereas any
replenishment of a depleted sulphate reservoir might
have been expected to return the total amount of tyramine conjugated to a mean value within the range observed in the control group.

Against this background, platelet PST M activity was
examined in a group of ten medication-resistant depressed patients, ten patient controls and ten staff controls (Bonham Carter et al.,1981b).The results for tyramine are shown in Figure 8 and it will be seen that no
difference in mean activity could be detected; neither
was there any difference in activity towards dopamine
or HMPG between the different groups (the samples for
monoamine oxidase assay mentioned above were obtained
during this same experiment).

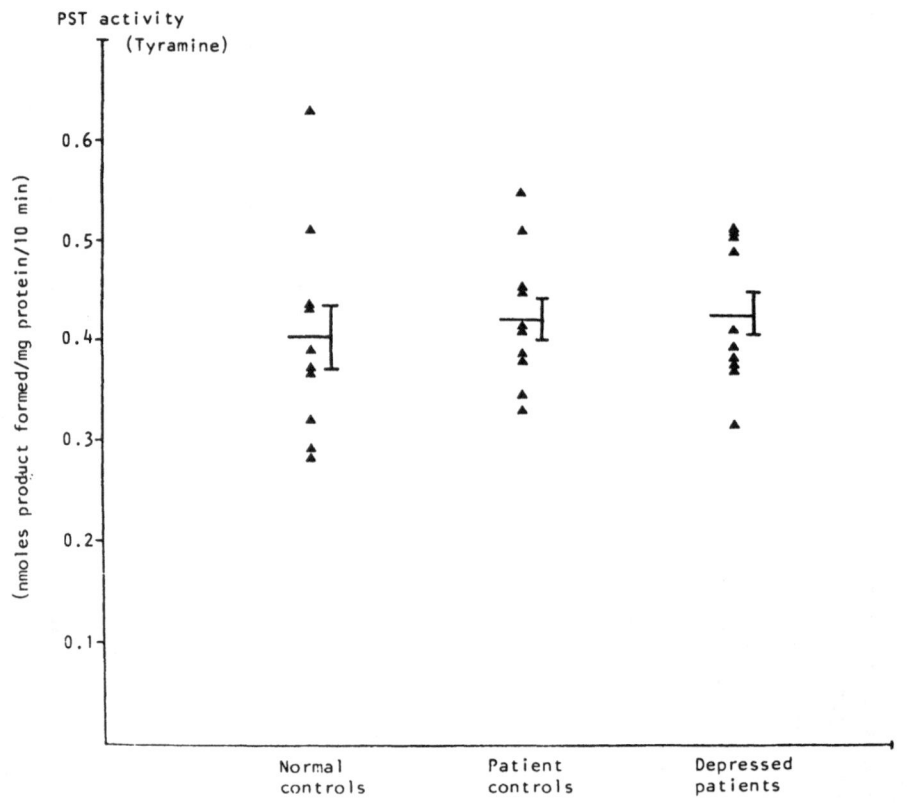

Figure 8
Platelet PST activity in control subjects, non-psychiatric patient
controls and severely depressed patients (Bonham Carter et al.,
1981b).

Thus, to judge from platelet activity, there is no evidence that the defect in tyramine conjugation we have observed stems from PST M deficiency. Platelet PST P activity was also examined in blood samples from the same three groups of patients using phenol as substrate and likewise there was no difference in mean activity between the groups.

It will be seen from the reaction sequence shown in the Introduction to this volume that the transfer of sulphate from PAPS to PST is only the final step in a complicated chain of chemical reactions which precede the process of sulphation.

Because of the obvious importance of ATP at two stages in the sequence, we thought it possible that a generalized deficiency of ATP might account for the tyramine conjugation deficit. Accordingly, whole blood ATP was measured in seven therapy-resistant depressed patients, six patient controls and seven staff controls (Figure 9) (Glover et al.,1981a). The values obtained were almost identical in each of the three groups so that no support could be obtained for an hypothesis involving a generalized deficiency of ATP in depressive illness.

Other possibilities still exist, including conceivably,a deficit in the synthesis of PAPS itself, for some reason yet to be defined. This problem is under active investigation but it should be noted that, although we are still ignorant of the cause of the tyramine conjugation deficit in depression, this has not prevented us using the finding as a trait marker. Bonham Carter et al. (1980b) recently showed, in a predictive study, that a significant increase in life-time incidence of depressive illness could be expected in a group of patients challenged with an oral tyramine load in late pregnancy who had poor conjugating ability compared with a group at the opposite end of the tyramine-conjugating scale (Figure 10).

PLATELET ACTIVITY IN PARKINSON'S DISEASE

Several years ago, Bonham Carter et al.(1974) found evidence of an increased baseline output of conjugated tyramine in untreated parkinsonians compared with controls. Later, Crowley and his colleagues (1978) reported an increased output of conjugated dopamine after L-dopa challenge in patients later shown to be particularly susceptible to drug-induced parkinsonism.

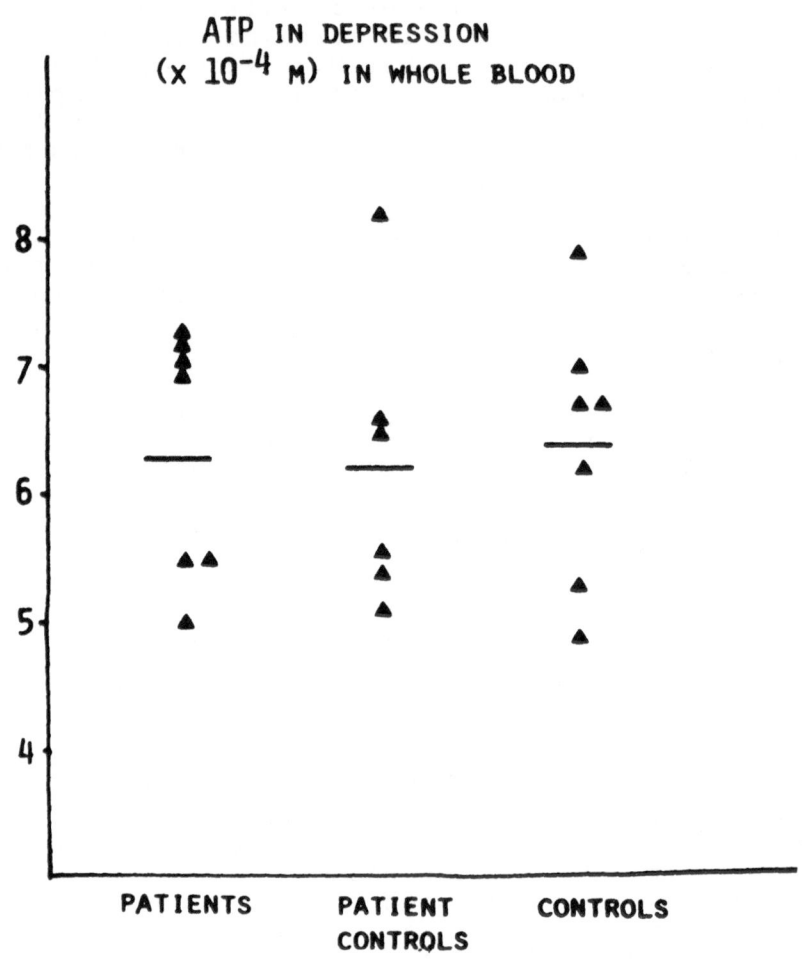

ATP IN DEPRESSION
(x 10^{-4} M) IN WHOLE BLOOD

PATIENTS   PATIENT   CONTROLS
              CONTROLS

Figure 9
Whole blood ATP levels in control subjects, non-psychiatric patient controls and severely depressed patients (Glover et al.,
1981a).

If an overactivity of PST were to be present in parkinsonian patients, it might exacerbate any pre-existing
dopamine deficiency by "siphoning off" dopamine otherwise available to act as transmitter.

Our first pilot study, comparing mean platelet PST
activity in a group of parkinsonian patients compared
with non-parkinsonian controls, demonstrated a significant ($p < 0.05$) increase in affected subjects (Glover
et al.,1981b). However, because almost all the parkin-

	HIGH TYRAMINE EXCRETORS	LOW TYRAMINE EXCRETORS
Depressed	4	11
Non-depressed	11	4

χ^2 (with Yates' correction) = 4.8

p = 0.014 (one-tailed)

Figure 10

Incidence of lifetime history of depression in individuals with high and low conjugated tyramine excretion after an oral tyramine load (100 mg) (Bonham Carter et al.,1980b).

sonians were taking L-dopa plus a peripheral decarboxylast inhibitor, the question arose as to whether the increase in enzyme activity was a drug response. In a larger, more recent study (Glover et al.,1981c), we have measured platelet PST activity in three different groups of parkinsonian patients, those treated with L-dopa alone, those treated with a lower dose of L-dopa combined with a peripheral decarboxylase inhibitor (Sinemet) and those not receiving dopa at all (some patients belonging to the latter group were receiving bromocriptine therapy). As shown in Fig.11, the group on L-dopa treatment alone manifested with a significantly higher mean activity than patients on L-dopa plus decarboxylase inhibitor (p < 0.05). This group in turn showed values significantly higher than parkinsonians not under treatment with L-dopa (p < 0.005, 2-tailed Student's t test). Values obtained for the parkinsonian patients untreated by L-dopa and non-parkinsonian controls were not significantly different. Thus we may assume that we are observing a drug-induced increase of PST in response to raised concentrations of dopa and dopamine - but this statement must be made with some slight qualifications: the untreated parkinsonians we observed were, inevitably, patients at an earlier stage of the disease than the treated groups, most of whom had been receiving L-dopa for substantial periods of time. Thus, we cannot

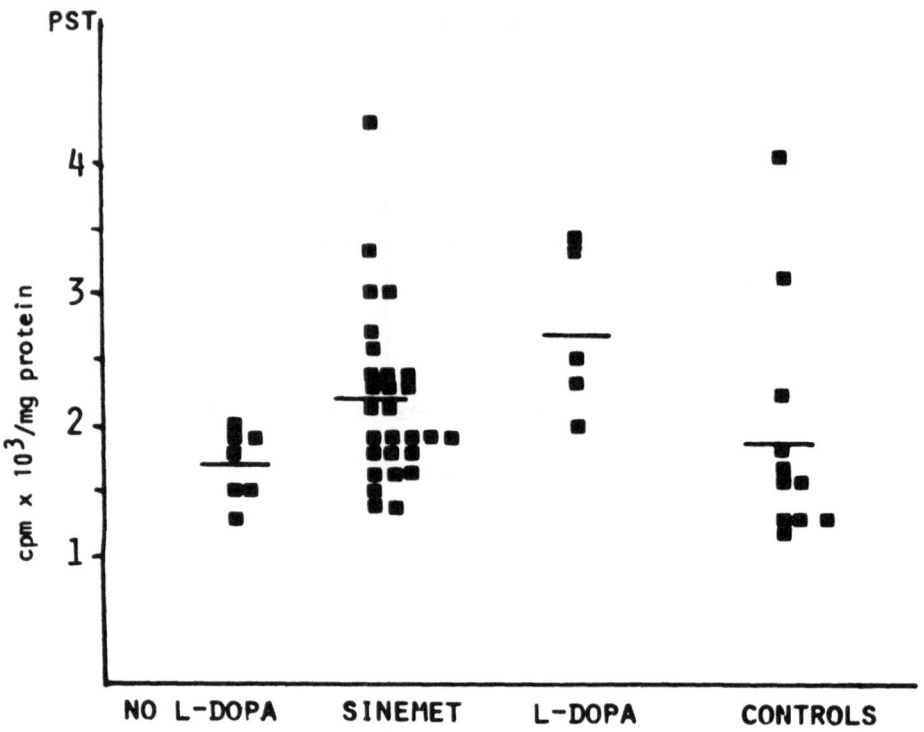

Figure 11
Effect of L-dopa and L-dopa with carbidopa (Sinemet) on platelet
PST activity (Glover et al.,1981c).

rule out the possibility that the rise in platelet PST
we have observed is a late manifestation of the parkinsonian process, although we consider this to be unlikely

Platelet MAO activity was slightly, but not significantly, lower in the parkinsonian patients than controls
but there was a significant correlation ($r = 0.39$; $p <
0.05$) between MAO and PST activities in the parkinson-
ian patients either on L-dopa alone or taking L-dopa
plus decarboxylase inhibitor (Figure 12) (Glover et al.,
1981c). This finding contrasts with a lack of correlation between these two enzymes in normal controls, dep-
ressives and headache patients. Thus, it seems likely
that, as with PST, MAO activity in platelets was induced
probably in this case by dopamine.

PLATELET PST ACTIVITY IN MIGRAINE

Because of the finding reported above coupling an

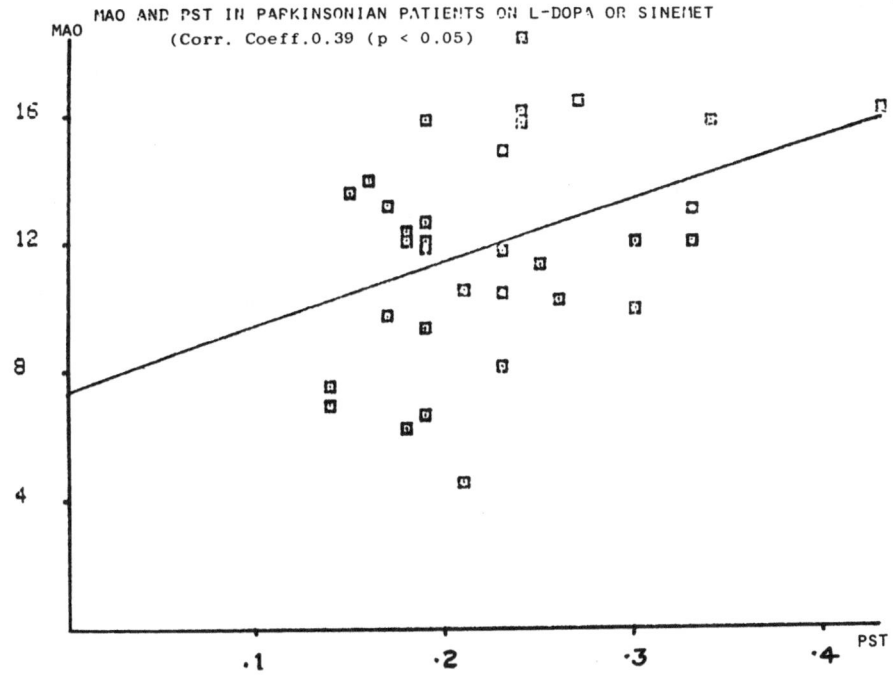

MAO AND PST IN PARKINSONIAN PATIENTS ON L-DOPA OR SINEMET
(Corr. Coeff.0.39 (p < 0.05)

Figure 12
Significant correlation between platelet MAO and PST activities
in parkinsonian patients on L-dopa or L-dopa with decarboxylase
inhibitor (Glover et al.,1981c).

increase of platelet PST activity with a rise in that of
MAO, at least after L-dopa treatment, and because there
have been extensive studies on platelet MAO activity in
migraine, it seemed important to measure PST activity
similarly in this group of diseases. We have recently
noted a significant decrease in platelet MAO activity
in this syndrome, outside a headache attack but in male
patients only (Glover et al.,1981d). The most highly
significant difference from the control group was observed in patients with cluster headache. We therefore compared platelet PST activity,outside a headache attack in
both male and female migraine sufferers (classified into three different groups, classical, migraine, common
migraine and cluster headache), with a group of normal
controls (Littlewood et al.,1981). (Figure 13). There
was no difference between male and female subjects. Patients with cluster headache did not differ significantly from control values - activity was slightly but not

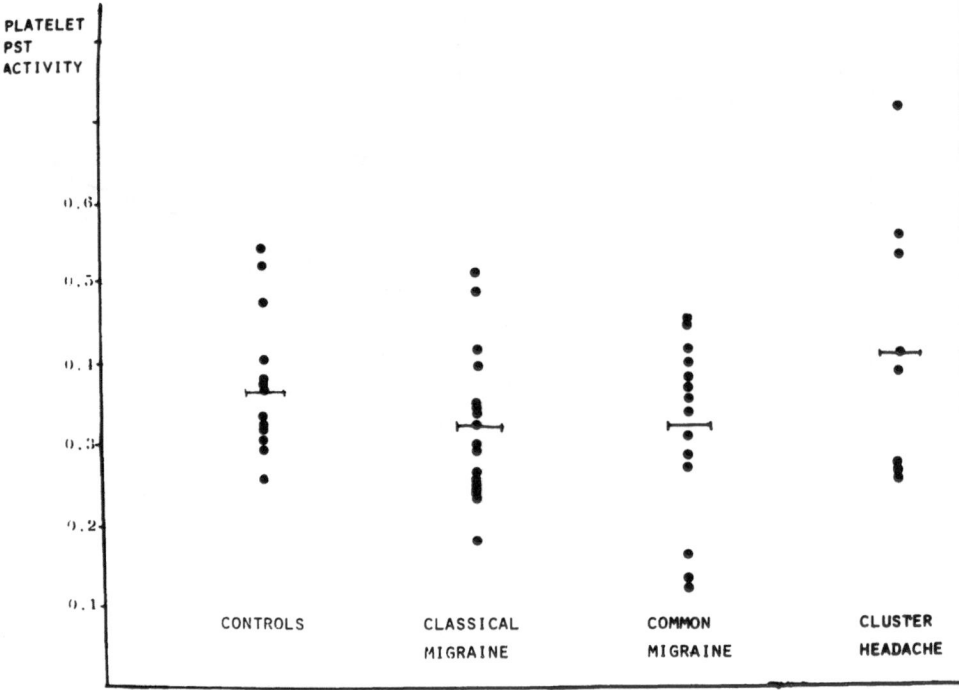

Figure 13
Platelet PST activity (nmoles product formed/mg protein/10 min)
in headache patients (Littlewood et al.,1981).

significantly higher. There was a small but significant
decrease in activity in classical plus common migraine
sufferers compared with controls, significant at the 1
in 20 level. Despite a considerable overlap between values in the migraine and other groups, there were a few
more individuals with values lower than any observed in
the control group and this finding caused the mean value
for the group as a whole to be reduced. Whether there
were any clinical differences between these outliers and
the rest of the group has not yet been investigated.

Although platelet MAO activity was also low in some
of these migraine patients, there was no correlation
whatsoever between activities of this enzyme and those
of PST (Figure 14). Whether a generalized deficiency of
either of these enzymes is associated with the sequence
of events leading to a migraine attack remains to be determined. It should be noted that the lack of correlation between the two enzyme activities does not necessarily provide evidence against the circulating humoral agent hypothesis of migraine (Sandler,1978), for these

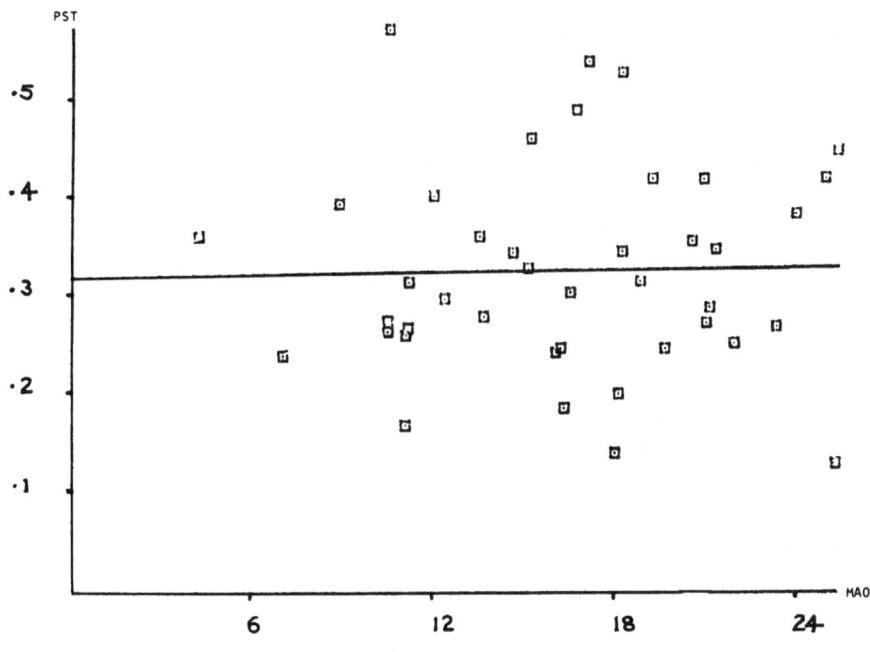

Figure 14
Lack of correlation between platelet MAO and PST activities in
headache patients (Littlewood et al.,1981).

enzyme changes were, in any case, observed outside an
attack. Because a separate decrease of platelet MAO
activity appears to occur during a migraine episode with
a return to values normal for that patient outside an
attack (Glover et al.,1977), it would be of interest to
determine whether a similar decrease, stemming perhaps
from the non-specific platelet damaging action of a
circulating humoral agent which is also responsible for
the headache and its vascular accompaniments (*c.f.*Sandler,1978), is present for PST also.

We would like to emphasize that migraine, like depressive illness and schizophrenia, is likely to be a
clinical mixture with a variety of different causes. Of
these syndromes, platelet PST activity in migraine obviously warrants further investigation but additional
studies in depressive illness seem unlikely to bear
fruit. Only schizophrenia has not so far been examined
from this viewpoint - but such a study is now in progress.

Anderson,R.J.and Weinshilboum,R.M.(1980). Phenolsulphotransferase in human tissue: radiochemical enzymatic
assay and biochemical properties. Clin.Chim.Acta
103, 79-90.

Anderson,R.J.,Weinshilboum,R.M.,Phillips,S.F.and Broughton,D.D.(1981). Human platelet phenol sulphotransferase: assay procedure, substrate and tissue correlations. Clin.Chim.Acta. In press.

Bonham Carter,S.M.,Glover,V.and Sandler,M.(1981a).Evi-
dence for separate control of two forms of human
platelet phenol sulphotransferase. Submitted for publication.

Bonham Carter,S.M.,Glover,V.,Sandler,M.,Gillman,P.K.and
Bridges,P.K.(1981b). Human platelet phenolsulphotransferase: characterisation for clinical studies and
activity range in depression. Submitted for publication.

Bonham Carter,S.M.,Goodwin,B.L.,Sandler,M.,Gillman,P.K.
and Bridges,P.K. (1980a).Decreased conjugated tyramine output in depression: the effect of oral L-cysteine. Br.J.Clin.Pharmac.,10, 305-308.

Bonham Carter,S.M.,Reveley,M.A.,Sandler,M.,Dewhurst,J.,
Little,B.C.,Hayworth,J. and Priest,R.G.(1980b). Decreased urinary output of conjugated tyramine is
associated with lifetime vulnerability to depressive
illness. Psychiat.Res.,3, 13-21.

Bonham Carter,S.,Sandler,M.,Goodwin,B.L.,Sepping,P.and
Bridges,P.K.(1978). Decreased urinary output of tyramine and its metabolites in depression. Brit.J.Psy-
chiat.,132,125-132.

Bonham Carter,S.M.,Youdim,M.B.H.,Sandler,M.,Hunter,K.R.
and Stern,G.M.(1974). Enhanced tyramine conjugation
in Parkinson's disease. Clin.Chim.Acta 51, 327-329.

Crowley,T.J.,Hoehn,M.M.,Rutledge,C.O.,Stallings,M.A.,
Heaton,R.K.,Sundell,S.and Stilson,D.(1978). Dopamine
excretion and vulnerability to drug-induced parkinsonism. Arch.Gen.Psychiat.,35, 97-104.

Eisenthal,R.and Cornish-Bowden,A.(1974). A new graphical
procedure for estimating enzyme kinetic parameters.
Biochem.J.,139, 715-720.

Foldes,A.and Meek,J.L.(1973). Rat brain phenolsulfotransferase - partial purification and some properties.
Biochim.Biophys.Acta 327, 365-367.

Glover,V.,Bonham Carter,S.M.,Sandler,M.,Kauffman,H.,   205
    Burnstock,G.,Gillman,P.K.and Bridges,P.K.(1981a).
    Whole blood ATP in depression. In preparation.

Glover,V.,Peatfield,R.,Zammit-Pace,R.,Littlewood,J.,
    Gawel,M.,Rose,F.C. and Sandler,M. (1981d). Platelet
    monoamine oxidase activity and headache. Submitted
    for publication.

Glover,V.,Sandler,M.,Grant,E.,Rose,F.C.,Orton,D.,Wilkin-
    son,M. and Stevens,D.(1977). Transitory decrease in
    platelet monoamine oxidase activity during migraine
    attacks. Lancet i, 391-393.

Glover,V.,Sandler,M.,Lees,A.and Stern,G.(1981c). Plate-
    let phenolsulphotransferase activity in Parkinson's
    disease. In preparation.

Glover,V.,Sandler,M.,Rein,G.and Stern,G. (1981b). Mono-
    amine oxidase and phenolsulphotransferase in Parkin-
    son's disease. In Progress in Parkinson's Disease
    (eds. F.Clifford Rose and R.Capildeo). Pitman Medi-
    cal, London. In press.

Hart,R.F.,Renskers,K.J.,Nelson,E.B.and Roth,J.A.(1979).
    Localization and characterization of phenol sulpho-
    transferase in human platelets. Life Sci., 24, 125-
    130.

Landowski,J.,Lysiak,W. and Angielski,S.(1975). Monoami-
    ne oxidase activity in blood platelets from patients
    with cyclophrenic depressive syndromes. Biochem.Med.,
    14, 347-354.

Levy,G.and Matsuzawa,T.(1967). Pharmacokinetics of sali-
    cylamide elimination in man. J.Pharmacol.Exp.Ther.,
    156, 285-293.

Littlewood,J.,Glover,V.,Peatfield,R.and Sandler,M.(1981).
    Platelet phenolsulphotransferase activity in headache
    patients. In preparation.

Mann,J.(1979). Altered platelet monoamine oxidase acti-
    vity in affective disorders. Psychol.Med.,9, 729-736.

Nies,A.,Robinson,D.S.,Harris,L.A.and Lamborn,K.R.(1974).
    Comparison of monoamine oxidase substrate activities
    in twins, schizophrenics, depressives, and controls.
    In Neuropsychopharmacology of Monoamines and their
    Regulatory Enzymes (ed.E.Usdin), Raven Press, New
    York, pp.59-70.

Nies,A.,Robinson,D.S.,Ravaris,C.L.and Davis,J.M.(1971).
    Amines and monoamine oxidase in relation to aging
    and depression in man. Psychosom.Med., 33, 470.

206   Rein,G.,Glover,V.and Sandler,M.(1981a). Multiple forms of human platelet phenolsulphotransferase; selective inhibition by dichloronitrophenol. Submitted for publication.

Rein,G.,Glover,V.and Sandler,M.(1981b). Sulphate conjugation of biologically active monoamines and their metabolites by human platelet phenolsulphotransferase. Clin.Chim.Acta. In press.

Reveley,M.A.,Glover,V.,Sandler,M.and Coppen,A.(1981). Increased platelet monoamine oxidase activity in affective disorders. Psychopharmacology. In press.

Sandler,M.(1978). Implications of the platelet monoamine oxidase deficit during migraine attacks. Res. Clin.Stud.Headache 6, 65-72.

Sandler,M.,Bonham Carter,S.,Cuthbert,M.F.and Pare,C.M.B. (1975). Is there an increase in monoamine oxidase activity in depressive illness? Lancet i, 1045-1049.

Author Index

Subject Index